AP® ENVIRONMENTAL SCIENCE
CRASH COURSE®

By Gayle N. Evans, M.Ed.

Research & Education Association
Visit our website at: www.rea.com

Research & Education Association
61 Ethel Road West
Piscataway, New Jersey 08854
E-mail: info@rea.com

AP® ENVIRONMENTAL SCIENCE CRASH COURSE®

Published 2017

Printed in the United States of America

Library of Congress Control Number 2011931952

ISBN-13: 978-0-7386-0931-7
ISBN-10: 0-7386-0931-5

Cover image: Spohn Matthieu/PhotoAlto Agency RF Collections/Getty Images

AP ENVIRONMENTAL SCIENCE CRASH COURSE
TABLE OF CONTENTS

PART I INTRODUCTION

PART II LIFE ON EARTH

PART III — HOW HUMANS USE AND CHANGE EARTH

PART IV — TYING IT ALL TOGETHER

PART V — TEST-TAKING STRATEGIES

ABOUT THIS BOOK

REA's *AP Environmental Science Crash Course* is designed for the last-minute studier or any AP student who wants a quick refresher on the course. The *Crash Course* is based on the latest changes to the AP Environmental Science course curriculum and exam.

Written by an expert who has been closely involved with the AP Environmental Science course since 2004, our easy-to-read format gives you a *Crash Course* in the major concepts and strategies in AP Environmental Science. The targeted review chapters will prepare you for the exam by focusing on important topics frequently seen on the AP Environmental Science exam.

Unlike other test preps, REA's *AP Environmental Science Crash Course* gives you a review specifically designed to zero in on the "big ideas" tested on the exam. Each chapter highlights the important themes and terms to keep in mind as you prepare.

Part I gives you the Keys for Success, so you can tackle the exam with confidence. It also gives you a review of basic math and science concepts as well as some key terms that you absolutely, positively must know. Part II presents essential information about Life on Earth from geological concepts, organisms, and atmospheric conditions.

Part III discusses how humans use and change the Earth—from pollution to energy resources to species extinction—the author concentrates on what you need to know for the exam.

When you're ready to prepare for the exam, REA's *AP Environmental Science Crash Course* will show you how to study efficiently and strategically, so you can boost your score!

To check your test readiness for the AP Environmental Science exam, either before or after studying this *Crash Course*, take REA's **FREE online practice exam**. To access your practice exam, visit the online REA Study Center at *www.rea.com/studycenter* and follow the on-screen instructions. This true-to-format test features automatic scoring, detailed explanations of all answers, and diagnostic score reporting that will help you identify your strengths and weaknesses so you'll be ready on exam day!

Good luck on your AP Environmental Science exam!

ABOUT OUR AUTHOR

Gayle Evans has a reputation among her students for her enthusiasm and passion for teaching about the environment. In the course of her career at Gainesville High School, in Gainesville, Florida, she has taught nearly every science course possible including AP Environmental Science, AP Biology, as well as Anatomy & Physiology and Physical Science.

Ms. Evans earned her B.A. in Biology from Mount Holyoke College in Massachusetts and her M.Ed. in Secondary Science Education from the University of Florida in Gainesville, Florida. She is also National Board Certified in Biological Sciences.

Ms. Evans would like to thank her students at GHS for making this book possible. Their excellent questions, spirited class discussions, and thousands of essays have provided her with the experience and the insight to be an effective, trusted, and successful educator.

ACKNOWLEDGMENTS

In addition to our author, we would like to thank Larry B. Kling, Vice President, Editorial, for his overall guidance, which brought this publication to completion; Pam Weston, Publisher, for setting the quality standards for production integrity and managing the publication to completion; and Diane Goldschmidt, Senior Editor, for editorial project management.

We would also like to extend special thanks to Stephen Everett, for technically reviewing the manuscript, Marianne L'Abbate for copy-editing, and Kathy Caratozzolo of Caragraphics for typesetting this edition.

PART I
INTRODUCTION

Six Keys for Success
on the AP Environmental
Science Exam

When you sit down to take the AP Environmental Science (APES) exam, you will find that this is a test that evaluates much more than your knowledge about environmental science concepts. In some ways, you are also being tested on how well you handle taking the test itself. The most important bit of advice I can give to you is to *relax*.

The APES exam is almost always given first thing on a Monday morning. This is great for you. You have the entire weekend to do some last-minute studying. Don't forget to take some time to de-stress from all that studying. Put your books aside every few hours and go outside. Look up at the sky and remember why you are learning about the environment. Breathe some fresh air and feel the breeze, stretch out your muscles. It is spring. The most beautiful time of the year! Let this inspire you when you get back to the books!

On the night before you take the exam, put your books away and go to sleep at a decent hour. Staying up all night studying is not going to help you, and it is more likely to prevent you from doing your best. Educational research has shown that sleep is an important part of processing new information. In order for all those things you are studying to stick in your long-term memory you need at least six to eight hours of sleep, so don't even think about pulling an all-nighter! Use this book to refresh your memory about all that you have learned this year, and you will be more than ready for test day!

1. Understanding the APES Scoring Scale

Most students are familiar with the grading scale that high school teachers use and assume the AP exam is graded on the same scale. This is not true. The good news is that the AP grading scale is much more forgiving. This is a challenging test. The College Board is aware of the fact that even the best students will not know the answer to every question.

As a result, the grading scale used is determined first by the performance of all students who take the test in a given year. If the test was more difficult than usual, the overall performance of students will be lower, and the grading scale will be adjusted downwards to compensate. In the same way, if the test is a little easier than usual, overall performance will be higher and the grading thresholds will rise.

Generally the AP graders determine the grade cut-off scores so that a little more than 50 percent of all students taking the exam score a 3 or higher. If you consider that in some schools, every student in an AP course is required to take the exam, regardless of motivation, the fact that you are reading this book already puts you at an advantage!

The APES exam is composed of 100 multiple-choice questions that make up 60% of your exam score, and four FRQs of ten points each that make up 40% of your score. When your exam is graded, your multiple-choice and free-response percentages are plugged into an equation to calculate a numerical value out of 150 points. Here is the grading breakdown from the 2008 released exam:

Score Range	AP Grade	Minimum percent right
102–150	5	68%
80–101	4	53%
67–79	3	45%
53–66	2	35%
0–52	1	0%

Look closely at these numbers and you will see that to pass the exam with a 3, you only need to accumulate 45% of all the available points. Answer just a few more questions correctly, and now you are in range to score a 4! You can do this!

Keep this in mind as you take the exam. If you come across a few questions that are totally unfamiliar, the knowledge that all you need is to get 68% of the questions right to get the highest possible score should keep you feeling confident that an awesome score is well within your reach!

2. **Understanding the APES Topic Outline**

The AP Topic Outline is published by the College Board as a guide for teachers to use in designing their APES course. Every question on the AP exam can be directly tied back to one of the topics on the course outline. If you have not already seen this outline, it is freely available from the AP Central website (*www.apcentral. collegeboard.com*). You can also download copies of released multiple-choice questions and free-response questions (FRQs) from past exams. I strongly suggest you use all of these materials as you prepare for the APES exam.

3. **Understanding the Importance of the Released Exams**

The College Board has released four complete exams to AP-certified teachers. The 1998, 2003, and 2008 exams that were administered to students have been made available to teachers to show what types of questions were used. In addition, in 2008, a practice exam was also made available. This exam has not actually been used with students, but was written as an example of the types of questions and topics students might see on a future exam. Every year, within a few days after the exam, the College Board shares the FRQs from that year's exam with teachers and students. This is a huge help in gaining insight into how to prepare for future exams.

The first thing that my students notice when we use these exams in class, is that there are certain types of questions that show up every time. For example, I have never seen an exam that did not include some type of radioactive half-life calculation, rule of 70 population growth problem or a question relating to an example of the tragedy of the commons. These are obviously important concepts to study. It is almost guaranteed that they will also show up on your exam.

Along the same lines, there are topics on the AP outline that have *never* shown up on a multiple-choice or free-response question on any released exam. These topics include the geologic time scale, specific information about cellular respiration, and the World Bank. This is not to say that they will not be on the exam in the future. A few years ago there was an FRQ about diseases like malaria, tuberculosis, and cholera that took everyone by surprise. No disease questions had ever been on a released exam! But, look closely at the course outline and right there in part III. "Population," section B. "Human Population," number 3, "Impacts of population growth"—the word *disease* is clearly a part of that list!

Taking all this into consideration, I have included the relevant information on every topic included in the course outline. Where a topic has never been used on a released exam or as an FRQ topic, I share that information with you. My advice is usually that you should be familiar with those topics, but do not spend too much time studying them, especially if your time is running short.

4. **Understanding the Importance of Key Topics**

An analysis of every question on the released and practice exams, as well as the FRQs from every year since the exam was developed in 1998, reveals some important clues about which topics are emphasized on the AP exam. Below is a table that, using the two most recently released exams, shows how the major topics correlate to the APES course outline.

Topic (followed by % given on the topic outline)	Percentage of multiple-choice questions found on 2008 exam	Percentage of multiple-choice questions found on 2003 exam	Percentage of FRQs (all years)
Basic Science	3	5	16.1
Earth Systems & Resources (10–15%)	13	10	5.5
The Living World (10–15%)	13	13	6.4
Population (10–15%)	13	12	6.6

Topic (followed by % given on the topic outline)	Percentage of multiple-choice questions found on 2008 exam	Percentage of multiple-choice questions found on 2003 exam	Percentage of FRQs (all years)
Land and Water Use (10–15%)	11	11	16.1
Energy Resources & Consumption (10–15%)	8	11	17.8
Pollution (25–30%)	24	20	20.7
Global Change (10–15%)	14	15	10.2

Examination of these data shows some important trends:

Although "Basic Science" concepts are not specifically listed on the APES course outline, they account for a significant proportion of the FRQ points. Just over 16% of all FRQ questions ever asked involved basic skills of data interpretation like reading data from a graph, performing simple calculations involving unit conversion, knowledge of experimental design, and the ability to graph data. If you need to review these skills, pay close attention to the practice problems in Chapter 2, "Basic Science and Math Concepts."

The topics under "Land and Water Use" and "Energy Resources & Consumption" are more likely to show up as FRQ questions than multiple-choice questions.

The opposite is true for "Pollution" and "Global Change." Both of these sets of concepts are more likely to be tested as multiple-choice questions.

If we look more closely at the specific questions in the multiple-choice section, we see that, although there is a lot of variation from year to year, some specific topics seem to be weighted a little more heavily. The following table shows the number of multiple-choice questions for the most commonly tested topics on the 2003 and 2008 released exams, and on the 2008 practice test.

Topic	2003 Multiple-Choice Questions	2008 Multiple-Choice Questions	Practice Test Multiple-Choice Questions
Population Dynamics	5	9	4
Pest Management	0	4	4
Fossil Fuels	2	2	6
Air Pollution	6	5	8
Solid Waste	4	3	3
Water Pollution	6	4	8
Stratospheric Ozone	4	4	2
Global Warming	7	7	7
TOTAL % of Multiple-Choice Questions	34%	38%	42%

So, looking at the table above, you would be wise to make sure you are well prepared to answer questions on the topics listed. These 8 topics generated a total of between 34 and 42 percent of all the multiple-choice questions on all three tests! A strong knowledge of these eight topics is going to put you well on your way towards earning that 4 or 5 on the AP exam!

5. **Realizing That the Multiple-Choice and FRQ Questions Are Drawn from the Same Pool of Information**

When you are studying for the APES exam you do not need to study separately for the two sections of the test. As you prepare for the multiple-choice questions, you are also preparing for the FRQs. All of the questions relate back to the topics in the APES topic outline. Your math practice will apply to both sections as well since there are regularly five to six calculation questions in the multiple-choice section, and you are guaranteed to have at least one calculation in the FRQ section.

6.   **Using Your *Crash Course* and Supplementary Materials to Build a Winning Strategy**

This *Crash Course* guide is specifically designed for your success. Although this guide is based on a thorough analysis of all of the released exam questions and the College Board topic outline, it also contains a wealth of advice based on my experiences with my students, and what has worked for them over the years.

For example, chapter 3 contains a list of the most important key terms you need to know to be fluent in the language of Environmental Science. Knowledge of the vocabulary of Environmental Science will help you interpret the questions as well as provide you with the factual information you'll need to answer the questions.

Chapters 4 through 23 contain detailed information relating the topics from the course outline to the past exam questions. Use these chapters to refresh your memory on all the concepts from the course or just focus on the specific chapters where you need more help.

Read Chapters 24 and 25 to tie all of the concepts together to form a "big picture" of the relationship of humans to the environment, and then read the last two chapters, 26 and 27, before you take the test for last-minute pointers on the strategies you can use to help you make the most of your knowledge and earn full credit for what you know.

If you're looking for more preparation before the exam, turn to REA's *AP Environmental Science All Access®* Book + Web + Mobile study system, which offers a comprehensive review book plus a suite of online assessments (chapter quizzes, mini-tests, full-length practice tests and e-flashcards), all designed to pinpoint your strengths and weaknesses and help focus your study for the exam.

Basic Science
and Math Concepts

The good news and the bad news about the AP Environmental Science (APES) exam is that there are always a number of questions that can be answered with no more than your basic science and math understanding and a good dose of common sense. If you are a junior or senior who has taken a number of AP courses already, this is a gift to you. These days, however, more and more students who take the AP Environmental Science course and exam are taking it as underclassmen, or as a first AP course. This chapter is for those of you who feel you could use a little brush-up on basic science and math skills. If you are a seasoned AP student, you may want to skim this chapter to be sure you remember some of the basics you may not have used in a while.

I. Science Review

The assumption of the writers of the AP Environmental Science exam is that students who are taking the exam have a solid science background that includes high school-level courses in biology, chemistry, and physics. The reality of the situation is that AP Environmental Science students are a very diverse mix of people. There is no "typical" APES student. The goal of this science review is to focus on specific skills and background knowledge that show up regularly on the AP exam, but are not part of the topic outline for the APES course. This will ensure that you are building your APES knowledge on a firm foundation.

A. Data Interpretation. It is guaranteed that you will be asked to work with scientific data on the exam. One or more of your free-response questions (FRQs) is guaranteed to include a map, table, graph, diagram or a word problem to solve.

1. World Map. Familiarize yourself with the world map. As you study things like climate and ocean currents, El Niño, human population, and distribution of the world's reserves of oil and coal, pull out a map and point to the places that you are studying. It is very common for questions like, "this is a location of high volcanic activity," or "this is the area of the world with the greatest coal reserves," to be given and associated with a world map like the one below. The map will be labeled with the letters A–E on different regions of the globe. A familiarity with the map will make these questions much easier.

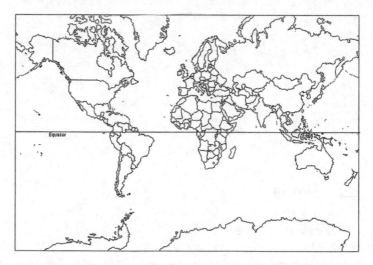

2. Graphs. You will see four to six graphs in the multiple-choice section of the exam and usually one or more on the FRQs. For the most part you will need to be able to pull data points from the lines on the graph, interpret relationships among different variables plotted on the same graph, and detect trends from the graphed data.

Math Practice

Use the climatogram below to answer the following data interpretation questions:

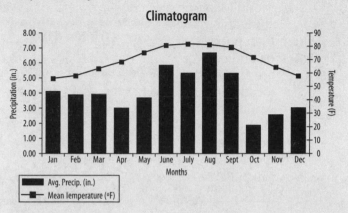

1. During what month is the rainfall the highest? How much rain fell in that month?

2. Which of these biomes does this climatogram most resemble: desert, subtropics, boreal forest?

3. What is the percentage of change in rainfall between January and October?

4. Is there a relationship between precipitation and temperature for this biome?

ANSWERS:

1. August, approx 6.5 inches.

2. Subtropics.

3. $\dfrac{4-2}{4} = \dfrac{2}{4} = 0.5$ $0.5 \times 100\% = 50\%$

4. Yes, they are proportional. In the months with highest temperature, precipitation is also highest. The months with lowest precipitation also tend to have the lowest temperatures.

3. Data Tables. You will see about three data tables in the multiple-choice section. Again, you'll need to interpret the

information to answer questions relating to the table. Below is a sample data table and typical APES exam questions:

Math Practice

World Coal Supply and Disposition, 2005 (Trillion BTU)				
Country	Production	Imports	Exports	Consumption
World Total	122,100	21,400	21,000	121,700
China	48,900	500	1,800	46,900
United States	23,100	800	1,300	22,800
Australia	8,500	0	5,900	2,500
India	7,800	1,100	40	8,600

1. Which country exports the greatest percentage of the coal it produces?

2. What percentage of the world's total production is represented by China, the United States, Australia, and India combined?

ANSWERS:

1. This question can best be answered by estimation. If you simplify the fractions and round your numbers, you will see that Australia is clearly the answer.

$$\text{China} = \frac{1,800}{48,900} \approx \frac{2}{50} = \frac{1}{25} \qquad \text{USA} = \frac{1,300}{23,100} \approx \frac{1}{20}$$

$$\text{Australia} = \frac{5,900}{8,500} \approx \frac{6}{9} = \frac{1}{3} \qquad \text{India} = \frac{40}{7,800} \approx \frac{4}{80} = \frac{1}{20}$$

2. $$\frac{48,900 + 23,100 + 8,500 + 7,800}{122,100} = \frac{88,300}{122,100} = 0.72$$

$$0.72 \times 100\% = 72\%$$

Test Tip

You can save a lot of time in the multiple-choice section if you estimate your answers instead of working out every problem! The answers are usually given in orders of magnitude (e.g., 1, 10, 100, 1000). You just have to be close to the answer to get it correct.

4. Diagrams. There is always at least one diagram to interpret in the multiple-choice section and one or more in the FRQs. I have included diagrams throughout this book that, based on past exams, are good candidates for inclusion in your exam. Study each of them and make sure you understand what it is meant to show.

B. Experimental Design. Although there is no single perfect way to design an experiment, all good experiments include several of the same features to ensure that the data collected is the best possible representation of reality.

1. Stating a Hypothesis. The hypothesis is a testable prediction. As the experimenter, what do you think is going to happen? The easiest way to state a hypothesis is as an "if, then" statement: "*If* bean plants are placed in soils of varying salinity, *then* those in the least salty soil will grow at a faster rate."

    ➤ There are two types of variables, the independent variable (what is changed), and the dependent variable (what is observed).

    ➤ A well-crafted hypothesis will make these variables clear. The "if " part of the statement addresses the independent variable (salinity of the soil); the "then" part of the statement addresses the dependent variable (growth rate).

2. Designing the Experiment. There are several things to consider when designing a good experiment.

    i. Control group. The purpose of the control group is to ensure that any changes you see in your test groups are caused by the independent variable.

        ➤ In the experiment above, the control group would be a group in which bean plants are grown in soil with no salt. If the plants grow well without salt, but poorly in the soil with salt, you can conclude that salt was the cause of the poor growth.

        ➤ If plants are grown in soils of varying salinity, but no control group is set, then a lack of growth could have been caused by the presence of salt or perhaps there was some other condition that led to poor growth (plant disease, lack of nitrogen in the soil, too much or not enough water, etc.).

ii. Only one variable is manipulated at a time. All groups should be identical except for the one independent variable.

➤ For example, the bean plants should all be the same variety, age, and approximate size. All plants should get the exact same amounts of water, sunlight, and soil nutrients. The only difference should be that salt is added to some groups and not to the control.

iii. Replication. To increase the validity of the data collected from an experiment, it should be able to be replicated. This means that it should be done more than one time to account for individual differences that may arise by chance.

➤ For example, maybe one plant just was not healthy and dies. In order to be certain its poor health was a result of its test conditions, you want to have several plants growing in that same set of conditions.

➤ There is no set number of replications that is always correct. Depending on the experiment, you may have as few as five or six replications per group or as many as hundreds of replications per group.

3. Making a Graph to Visualize the Results. There may be an FRQ in which you are given a set of data and a blank grid and are asked to set up a graph and plot the data. To ensure you don't lose any points, be sure to consider all of the following:

i. Set up your axes so that the independent variable is always on the horizontal (*x*) axis, and the dependent variable is on the vertical (*y*) axis.

ii. Make sure you include a descriptive label, with units, for each axis.

➤ For the above experiment, the *x* axis should be something like "Salinity of soil (mg/ml)."

➤ The y-axis could be "Growth of bean plants (mm)."

iii. Determine an appropriate scale for each axis so that you use at least 75% of the grid provided, and keep a consistent scale interval (count by .1, .2, .3, etc., or 10, 20, 30…)

When asked to make a graph, the dimensions of the grid given will suggest what your intervals should be. For example, if counting by tens is best on your y-axis, then the number of squares given for that axis will be such that if you count by tens, you will have exactly the right number of squares. Don't panic, however, if you don't end up with an exact count. There is always a range of scales that are accepted as long as they are at consistent intervals. Include all of the data and take up at least 75% of the grid.

iv. Line or bar graph?

➤ A line graph is used to show continuous data, such as how something changes over time. For our bean experiment, a line graph would be the best choice.

➤ Bar graphs are used to display discrete data. For example, if we were to count cars in a parking lot and graph them by color, each bar on the graph would represent all of the cars of a single color.

C. Feedback Loops. Every system in nature is in a nearly constant state of change. In most cases, these changes may be small adjustments that keep the system in balance. In other cases the changes are greater and throw the system out of balance. These changes are referred to as feedback loops.

1. A *negative feedback loop* is a state of adjustment that helps keep a system stable. Think of the heating and air conditioning systems in a home. The desired temperature for the house is set and when it gets too warm, the air conditioner kicks on to cool the house; when the desired temperature

is reached, it turns off. If it gets too cold in the house, the furnace comes on to warm the house back to the desired temperature, then turns off.

2. A *positive feedback loop* happens when changing something about a system sets up a situation where that change is amplified further and further until some ultimate end point is reached. Imagine a Petri dish full of bacteria. A small number of bacteria are put in the dish to start, and every couple of hours the bacteria population doubles in size. Every time the population doubles, there are twice as many bacteria to reproduce, so the rate of population growth increases and increases until eventually the dish is so crowded with bacteria that the food runs out, causing a population crash in which most of the bacteria die off.

3. Positive feedback loops are often called vicious cycles because once they begin, it is difficult to stop or reverse them.

4. The terms "positive" and "negative" refer to the direction or amount of overall change in the system. Negative feedback loops contribute to overall stability (no change), whereas positive feedback loops lead to increasing change in a system.

Test Tip

Many students find this concept of feedback loops confusing because they think that "positive" must be good and "negative" must be bad. However, in most cases, the reverse is true when thinking about feedback loops.

D. The Relationship of Temperature to Density and Volume. All matter in the universe is made up of atoms that combine to make molecules. The states of matter—solid, liquid, and gas—are determined by how energized the molecules of a substance become.

1. For example, if we take liquid water and place it in a very cold environment, heat is lost from the water and its molecules move less and less until they "freeze" into the crystal lattice structure we call "ice."

2. Warming up this ice adds energy to its molecules. When they acquire enough energy to move around in three dimensions—sliding over and around one another—the ice melts to become liquid water.

3. If we add heat to this water, eventually its molecules will become so energized that they spread out enough, and have such violent collisions, that they begin to leap out of the water and become water vapor.

4. In most cases, as a substance changes from solid to liquid to gas, the molecules spread further and further apart as energy is added. Thus, solids tend to have the most molecules per unit of volume, liquids have fewer molecules in that same volume, and gases even fewer molecules in that volume.

5. *Density* is defined as mass per unit of volume. Since molecules have mass, more molecules per unit of volume = greater density.

6. For most substances, solids (which exist at lower temperatures) have the highest density, followed by liquids (which exist at intermediate temperatures), and gases (which exist at higher temperatures) have the lowest density.

7. The relationship between temperature and density applies to a substance even if the differences are not great enough to lead to a change in phase. For example, one of the key concepts driving weather patterns is the fact that hot air tends to rise while cold air tends to sink towards the ground.

II. Math Review

The APES exam is guaranteed to have math questions in both the multiple-choice and free-response sections of the exam. Expect about 2 or 3 calculations in the multiple-choice questions and be prepared to find that at least one of your four FRQs will be primarily based on data interpretation and mathematic calculation. You won't be allowed to use a calculator. Before you frown, realize that this is good news. Without a calculator, the AP exam writers don't expect you to do anything too complex. A basic knowledge of algebra, and a little common sense, is all you really need to rock these questions!

Some people see this as bad news. These days, it is rare to not be able to use a calculator on an exam. Because we use them so often, many of us feel uncomfortable doing basic addition, sub-traction, multiplication, and division without a calculator. The best way to overcome this fear is to break out the pencil and spend some time practicing. Even if you have to "Google" long-division to refresh your memory, a basic math skill set that will serve you well on the APES exam and in life. Think about how great it will be to always know exactly how much to tip the waiter!

Throughout this guide, you will find math practice problems. These practice problems will give you a chance to see exactly what types of math questions you will encounter on the exam. For now, let's review some basic skills and tips that will give you the tools you need to tackle any type of question you might encounter.

A. When given a question with several numbers, don't panic!

1. First, look closely at each number and its units. If the num-bers are written into a long paragraph, pull them out and make a list to the side of each number and its unit.

2. Next, read the question to find out exactly what is wanted in the answer. If you know the unit of the answer, you can work backwards to figure out how to set up your numbers to cancel out the units you don't need and be left with only the units you do need. This process is called dimensional analysis and it can make the most complicated question seem truly simple.

3. Let's look at an example:

    The only time you are likely to see an adult sea turtle on the beach is during nesting season. Every two years, adult Loggerhead sea turtles find mates and the females come up on to the sandy beaches to dig nests and lay their eggs. A female Loggerhead will lay about 100 eggs per nest. In a nesting year, she will lay about five nests. Of those eggs, about 80% emerge as hatchlings. It is estimated that only 1 in every 1000 hatchlings survives to become an adult sea turtle.

If Loggerhead sea turtles reach sexual maturity at age 25, how many years must a female loggerhead live to produce enough adult turtles in the next generation to replace both her and her mate?

Begin by listing all of the numbers and units:

Nests every 2 years with 5 nests/nesting year

100 eggs/nest → 80% hatchlings

1/1000 hatchlings → adulthood

25 years to adulthood

Then notice that you want your answer in years. Then think, "If I need two adult turtles, and only 1/1000 hatchlings lives to adulthood, then I need 2000 hatchlings." Now you know that you need to find how many years it takes to produce 2000 hatchlings.

Here's one way to work it out:

$$\frac{5 \text{ nests}}{2 \text{ years}} \times \frac{80 \text{ hatchlings}}{\text{nest}} = \frac{200 \text{ hatchlings}}{\text{year}}$$

Now, how many years does it take to produce the 2000 hatchlings needed to get two adult turtles?

$$2000 \text{ hatchlings} \times \frac{1 \text{ year}}{200 \text{ hatchlings}} = 10 \text{ years}$$

(Notice that in the equation above, 1 year/200 hatchlings was flipped to cancel hatchlings and end up with years as the unit.)

So in just a few easy steps we have used dimensional analysis to keep our units straight and get the correct answer! You will find that most of the calculation-based questions on the APES exam will work the same way.

Test Tip

Be sure you isolate the numbers out of the paragraph, keep track of your units, and set up your equation to cancel out all the units you don't want to leave behind in the answer! It is easy to get lazy or rushed and start dropping the units from your rough calculations, but, trust me, this will only cause problems down the line.

B. After you complete your calculations, do a common sense check. Ask yourself, "Does this answer make sense given my knowledge of the world and my common sense?" For example, if your answer above had been 350 years old because of an extra stray zero, you would stop and think, "Hmmm, if Loggerhead sea turtles have to live 350 years just to replace themselves, how has this species survived for millions of years?" The answer is, of course, "Impossible!" If this happens to you, simply go back and re-check your work to find your mistake. Did you multiply when you needed to divide (this is where writing out units and canceling out is helpful)? Did you add or drop a zero or two between steps? These are all common errors that we tend to make when we are rushing or under stress.

C. Calculating percent increase or decrease shows up in many situations. For problems like this, use the following equation:

$$\frac{\text{change in amount}}{\text{original amount}} \times 100\%$$

Subtract the smaller number from the larger number to find the change in amount (this must always be a positive number), then divide by the original amount.

Here's an example.

A scientist monitoring a population of water beetles noticed that in 2001 the population was 500 beetles. Ten years later, she counts only 300 water beetles in the population. What is the percent decrease of this population?

$$\frac{500-300}{500} = \frac{200}{500} = 0.4$$
$$0.4 \times 100\% = 40\% \text{ decrease}$$

Say the population went from 500 water beetles in 2001 to 600 in ten years. What is the percent increase of the population?

$$\frac{600-500}{500} = \frac{100}{500} = 0.2$$
$$0.2 \times 100\% = 20\% \text{ increase}$$

D. Exponential notation is a great way to simplify numbers that are either very large or very small. Because you cannot use a calculator, an understanding of how to use exponential notation will be very handy on the APES exam. You may want to review exponents in a math or science book, but basically, here's all you need to know for the exam:

You can simplify a very large number like this:

4 million = 4,000,000 = 4×10^6
(the 10^6 represents the six zeros in 4 million)

Likewise, 4 billion = 4,000,000,000 = 4×10^9

Test Tip

> It's a good idea to memorize the exponential notations 10^6 = million, 10^9 = billion, and 10^{12} = trillion since these are the most commonly used units in APES for large numbers.

To use exponential notation with very small numbers, determine your exponent by counting how many places to the right you have to move the decimal point to get a whole number. For example:

4 millionths = 0.000,004 = 4×10^{-6}

10^{-6} is the correct exponent because you had to move the decimal over six places to the right to get 4. Using exponential notation is probably most useful when multiplying or dividing very large or very small numbers.

Here's an example of a situation you might see:

> *The world's population is nearly 7 billion people. There are about 300 million people in the United States of America. Approximately what percentage of the world's population is represented by the United States? (Round to the nearest whole number.)*

$$\frac{300\ million}{7\ billion} = \frac{300 \times 10^6}{7 \times 10^9} = \frac{30 \times 10^7}{7 \times 10^9} \cong 4 \times 10^{-2} = 0.04$$
$$0.04 \times 100\% = 4\%$$

So, what happened here? First, convert the millions and billions to exponential notation, then divide 30 by seven to get, roughly, four. To divide exponents, subtract the bottom exponent from the top exponent (7 − 9 = −2). After that, convert the number back to a decimal and multiply by 100% to convert the answer to a percentage.

If asked to multiply a number in exponential notation, simply add the exponents:

$$\left(4\times10^{6}\right)\times\left(3\times10^{6}\right)=12\times10^{12}$$

Finally, if adding or subtracting numbers in exponential notation, convert them to the same exponent first. If you wanted to add 67 million to 4 billion, here's the calculation:

$$\left(67\times10^{6}\right)+\left(4\times10^{9}\right)=\left(67\times10^{6}\right)+\left(4000\times10^{6}\right)$$
$$=4067\times10^{6}$$

REFERENCES

World Map: *http://www.colby.edu/geology/gifs/worldmap.gif*

Data for Climatogram: *http://www.weather.com/weather/ wxclimatology/monthly/graph/32605*

World Coal Supply Data: *http://www.eia.doe.gov/iea/coal.html*

Key Terms

 I. **Earth Systems and Resources**

A. Geologic Time Scale versus Human Scale

1. Current scientific evidence places the Earth's age at more than 4 billion years. When we think of the ways the Earth changes in geologic time, those changes are occurring over the course of hundreds of thousands of years, at a rate that would be undetectable in human time.

2. Human time runs on a much shorter scale. Each human generation is only about 30 years long. The major changes that humans have brought about on the Earth have all occurred within the last few thousand years. Most people think in terms of the next hundred years.

3. Many people have difficulty grasping the large spans of time between geologic events, which creates problems under-standing plate tectonics, evolution, and other scientific phenomena that occur over very long spans of time.

B. Earth's Spheres

1. When studying the Earth, we often divide Earth's activities into four spheres:

   i. Lithosphere

   The *lithosphere* is composed of the solid, outermost layer of the Earth. It includes the crust and upper mantle. The lithosphere is mostly rock and includes soils, silts, and sediments. It makes up the ocean floor, mountain ranges, and everything we think of as ground.

ii. Atmosphere

The gases that surround us and make life on Earth possible are known as the *atmosphere*. The air we breathe and our weather systems occur in the atmospheric layer closest to the ground called the *troposphere*. The troposphere is made up mostly of nitrogen gas (about 78 percent) and oxygen (21 percent) but also includes water vapor, carbon dioxide, and pollutants such as methane and oxides of nitrogen and sulfur. Above the troposphere, about 11 miles above sea level, is the *stratosphere*, which contains the ozone layer that protects Earth from too much UV radiation. Above the stratosphere are layers called the *mesosphere, thermosphere*, and *exosphere*.

iii. Hydrosphere

All of the water on Earth is collectively called the *hydrosphere*. This includes the oceans, which cover about 70 percent of Earth's surface and contain about 97 percent of Earth's water supply, as well freshwater lakes, streams, rivers, and underground aquifers. Less than 1 percent of the water on Earth is available for human use, such as drinking and irrigation. Of that water, most is stored as groundwater in underground aquifers.

iv. Biosphere

➤ The *biosphere* encompasses all life on Earth. The biosphere spans across all three of the other spheres because there is life in soil, water, and air.

➤ *Biomass* is organic material made from living things, like wood from trees, organic content in soils, or peat harvested from bogs. Biomass materials may be burned as a source of energy.

C. Plate Tectonics

1. *Plate tectonics* is the scientific theory that explains the formation of deep ocean mountain ranges and trenches and the occurrence of volcanoes and earthquakes. The theory states

that the Earth's crust is broken into several large plates that move slowly over a hot layer of the Earth called the *aesthenosphere*. The movement of these plates in relation to one another causes several geological conditions, especially along plate boundaries. There are three typs of plate boundaries: the convergent boundary, the divergent boundary, and the transform boundary.

i.   Convergent Boundary

➤ When two plates move toward one another, they eventually collide and one plate will move on top of the other, creating a subduction zone. This often results in mountain range formation and volcanic activity as seen around the so-called Ring of Fire in the North Pacific. In most cases, the overriding plates are continental crust that rides over the top of the more dense oceanic crust.

➤ Deep trenches, like the Mariana trench in the Pacific Ocean, also result from convergent boundaries.

➤ The Himalaya mountain range was formed by a convergent boundary between the Indian and Eurasian plates.

ii.   Divergent Boundary

➤ When two plates move away from one another, a gap forms and magma from the mantle rises through the gap to form broad mountain ranges and rift valleys.

➤ The African Rift Valley running through Ethiopia, Uganda, Kenya, and Tanzania is caused by a divergent boundary.

➤ The most notable divergent boundary mountain range is the Mid-Atlantic Ridge, which runs along the ocean floor from the Arctic Ocean to near the southern tip of Africa.

  iii. Transform Boundary

  ➤ When two plates slip along one another side by side, this often forms a transform fault and results in earthquakes.

  ➤ The most notable transform fault is the San Andreas Fault in southern California.

II. The Living World

A. Biotic and Abiotic

  1. When we discuss the interaction of living organisms with the environment, we often use the term *biotic* to mean anything that is or was living.

  i. Examples of biotic components of an ecosystem include any living plants, animals, fungi, and bacteria, as well as leaf litter, animal wastes, and remains of dead organisms.

  ii. A biotic factor is anything that results from interactions with living things. For example, parasitism, a bacterial infection, and even competition are all biotic factors.

  2. *Abiotic* means anything that is nonliving.

  i. Examples of abiotic components include air, water, rocks, and anything else in an environment that is not a direct product of a living thing.

  ii. An abiotic factor is anything nonliving that affects life, such as the weather or climate, including temperature, rainfall, and sunlight.

  iii. Other abiotic factors include soil texture and moisture; atmospheric pressure; and the presence of chemicals such as oxygen, acids, and nitrogen-containing compounds in the air, water, and soil.

B. Ecosystem

  1. An *ecosystem* is defined as the interaction of all the living organisms in an area with their nonliving environment.

2. Ecosystems on land are called *terrestrial ecosystems,* while those in the water are called *aquatic ecosystems.*

C. Species

1. *Species* is defined as a group of organisms that share similar physical and behavioral traits and that can interbreed to produce fertile offspring.

    i. Keystone Species

        ➤ In many ecosystems, one or more species are essential to the maintenance of the ecosystem. If a keystone species is removed from its ecosystem, the whole system is at risk of total collapse.

        ➤ Top predators, like sea otters and coyotes, are common keystone species.

        ➤ Environmental engineers are a type of keystone species that physically manipulate the environment in a way that makes their ecosystem possible.

        ➤ African elephants are an example—they uproot small tree saplings in their quest to find food, thus keeping the forest from encroaching into the grassland.

    ii. Foundational Species

        ➤ Foundational species are the autotrophs found in large numbers at the base of an ecosystem's food web.

    iii. Indicator Species

        ➤ Indicator species are especially sensitive to changes in the environment.

        ➤ Many frog species are indicators of water pollution because their permeable skin makes them especially sensitive to toxins.

        ➤ Mayflies are an indicator of high water quality in streams and rivers. They are also sensitive to pollutants, so finding mayflies and their larvae in an ecosystem indicates high water quality.

D. Habitat Versus Niche

1. Habitat and niche are related concepts that are often confused.

   ➤ The *habitat* of an organism is where it lives, while a species' *niche* describes its role within the ecosystem.

   ➤ A shark's habitat may be described as the open ocean or coral reef, while its niche would be predatory fish.

E. Community

1. A *community* is the assemblage of all organisms living and interacting in a particular area.

   ➤ A community includes all of the different species of plants, animals, fungi, bacteria, and protists.

   ➤ Add the nonliving environment to a community and you have an ecosystem.

F. Trophic Levels

1. *Trophic levels* are also known as feeding levels.

   i.   The first trophic level in any ecosystem are the producers—autotrophs that make their own food. In most cases, autotrophs are photosynthetic plants or chemosynthetic bacteria.

   ii. The next trophic level is the primary or first-degree consumers—herbivores (plant eaters) that feed on producers.

   iii. Secondary or second-degree consumers are the omnivores and carnivores that feed on primary consumers. This may continue to tertiary (third-degree) consumers feeding on secondary consumers and quaternary (fourth-degree) consumers feeding on tertiary consumers.

   iv. Most ecosystems have no more than four or five trophic levels due to the fact that only 10 percent of the energy available at one trophic level is transferred to the next highest level.

G. Biome

1. A *biome* is a large region of the Earth characterized by a distinct set of climate conditions.

2. Biomes are determined primarily by temperature and precipitation.

3. Major biomes of the Earth include forests, grasslands, deserts, wetlands, and estuaries.

   i.   Forests

   ➤ Forests are characterized by the growth of mature trees and a closed canopy. For a forest to be possible, rainfall must be high enough to support growth of large trees. As a result, forests tend to be the biomes with the highest precipitation.

   ➤ Forest types are largely determined by the seasonal temperature range and the length of the growing season.

   — *Tropical forests* tend to be evergreen and are warm year-round with a large amount of rainfall.

   — *Temperate forests* are usually deciduous (shed leaves in the winter), and have four equal seasons (spring, summer, fall, and winter), a wide range of temperatures, and moderate rainfall.

   — *Boreal forests* (also called Taiga) are evergreen forests with short growing seasons; long winters; and moderate to high precipitation in the form of rain, snow, and dew.

   ii. Grasslands

   ➤ Grasslands are characterized by a lack of trees and an abundant growth of grasses and other herbaceous (nonwoody) plants.

   ➤ Rainfall in grasslands is generally too low to support the growth of large woody plants, including trees.

➤ Grassland types are also determined largely on seasonal temperature ranges.

— *Tropical grasslands* like the African savannah tend to be warm year-round, and the growing season is largely determined by seasonal droughts.

— *Temperate grasslands,* like the prairies of the central United States, the Veldts of South Africa, the steppes of Russia, and the South American Pampas, tend to have cold winters and short growing seasons. Because the grasses die off each winter and have several months to decompose before the next growing season, the soils of temperate grasslands are extremely rich with organic material and tend to make excellent agricultural land for growing crops and grazing livestock.

iii. Deserts

➤ Deserts are the driest biomes, usually with so little rainfall that the diversity of life tends to be low.

➤ Generally deserts are characterized as either hot or cold.

— Hot deserts are what we normally think of as desert. They are extremely hot and dry with sandy soils.

— Cold deserts include the polar deserts, where it tends to be very cold with low precipitation and permafrost or ice cover for most of the year.

iv. Wetlands

➤ A wetland is an area in which the soils are flooded with water during part of its natural cycle. Water plays an important role in determining the plant and animal life of an ecosystem.

➤ Swamps and marshes are wetlands that have standing or flowing water year-round.

➤ Bogs and fens usually have saturated soils and occasional standing water, but they generally have less water and seasonal dryness.

   v. Estuaries

➤ Estuaries are areas where saltwater mixes with freshwater.

➤ Often found at the mouth of a river where it flows into the ocean, estuaries are known for high biodiversity.

➤ Their brackish (slightly salty) water makes estuaries a valuable habitat for shellfish, birds, fish, and other types of wildlife to reproduce.

H. Reservoirs

1. A reservoir is a zone of storage or containment.

2. When studying the biogeochemical cycles, like the carbon, nitrogen, water, phosphorus, and sulfur cycles, an understanding of reservoirs is important.

3. For example, the greatest reservoir for carbon is the calcium carbonate in limestone deposits, but fossil fuels are also a significant carbon reservoir.

4. A reservoir can also be a large human-made lake of water that accumulates behind a large dam. These reservoirs are often used to supply water to nearby urban areas.

 III. **Population**

A. Anthropogenic

1. *Anthropogenic* refers to any change in the Earth's systems caused by human activities.

   i. Most pollution is anthropogenic;

   ii. Cultural eutrophication is an anthropogenic increase of nutrients in an aquatic ecosystem.

  iii. Deforestation for agriculture and urbanization are also examples of anthropogenic changes.

 2. The debate over global climate change is whether the rising temperatures we have seen recently are anthropogenic or natural. As human populations continue to increase, we are having greater and greater anthropogenic effects on the Earth's systems.

B. Per Capita

 1. *Per capita* means "per person."

 2. Per capita income for a nation is a measure of how much, on average, each person in that nation earns.

IV. Land and Water Use

A. Environmental Impacts, Consequences, or Implications

 1. An *environmental impact* is some change in the environment that occurs as a result of some type of activity.

  ➤ Every living organism interacts with the environment, resulting in environmental impacts.

  ➤ Possible environmental impacts of humans burning fossil fuels include habitat destruction and water acidification from the mining process, and air pollution from the burning process.

Often the terms consequences *and* implications *are used interchangeably in APES exam questions. This is because essentially they mean the same thing: how are our human activities affecting the environment?*

B. Arable Land

 1. *Arable land* is any land suitable for agriculture.

  ➤ One ever-increasing problem will be our limited amount of arable land.

➤ As populations grow, we may not have enough arable land to grow the crops needed to feed Earth's people.

C. Irrigation

1. Most crops we grow for food need a steady supply of water. To increase the reliability of our harvests and to produce more food per acre, we often supplement the natural rainwater that falls with additional water taken from groundwater, lakes, or rivers. This process of watering crops is called *irrigation*.

   i. Salinization

      ➤ Salinization is a common problem associated with irrigation. While rainwater contains few impurities, water that we take from land-based sources will always have a small amount of dissolved salts resulting from contact with sediments and soils. When we irrigate crops, these salts build up in the soil surrounding the plants and gradually make it salty. This is a particular problem in arid areas with little rainfall because rainfall works to wash away these accumulated salts.

   ii. Desalinization (or Desalination)

      ➤ One solution is desalinization, which dissolves the accumulated salts in water by flooding the affected soil and draining the salty water away.

   iii. Water Diversion

      ➤ One of the original ways humans irrigated crops was to divert water from a lake or stream using a series of canals that directed the flow of water toward crops. Water diversion projects are still in use today all over the world, only now the scale is much larger.

      ➤ The shrinking Aral Sea in Russia is a famous example of the problems that can result from large-scale water diversion projects.

D. Extraction

　　1. The term *extraction* refers to mining, drilling, and removing trees from a forest. Any time we remove a resource from the Earth, it is referred to as extraction.

V. Energy Resources and Consumption

A. Laws of Thermodynamics

　　1. The laws of thermodynamics describe how energy behaves in Earth's systems. Make sure you understand both the first and second laws and how they apply to energy.

　　　　i.　First Law of Thermodynamics

　　　　➤ The law of conservation of energy states that energy can neither be created nor destroyed. As a result, all the energy in the universe is held constant.

　　　　➤ Energy may be transferred from one object to another in many forms.

　　　　ii.　Second Law of Thermodynamics

　　　　➤ The law of entropy states that energy always moves from a more concentrated state to a less concentrated state. This is why energy moves from hot objects toward cool objects (and never in the opposite direction).

　　　　➤ As a result, all natural processes involving energy transfer are linear and cannot be reversed.

　　　　➤ The sun is a one-way source of energy for the Earth. Energy moves through ecosystems from the producers at the base to the top predators at the apex.

　　　　➤ The second law of thermodynamics also explains why 90 percent of energy is lost at each step in the food chain, and why your car needs to be refilled with gas to keep going. At each conversion from one form to another, some of the energy is degraded, or lost to heat and/or friction.

iii. Efficiency

➤ *Efficiency* is a measurement of how much energy used in a reaction is converted to useful activity and how much is wasted. A highly efficient reaction wastes very little energy.

➤ The second law of thermodynamics states that it is not possible to have a 100 percent efficient energy conversion.

B. Potential and Kinetic Energy

1. *Potential energy* is stored energy that can be unleashed to do useful work.

2. *Kinetic energy* is energy in motion.

➤ A lump of coal represents potential energy. The energy stored in the carbon and hydrogen bonds in that lump of coal can be converted to kinetic energy by burning the coal.

VI. Pollution

A. *Pollution* is defined as anything added into the environment that may cause harm.

1. Air Pollution

i. *Air pollution* is defined as any gases, particulate matter (smoke and dust), and biological entities (pollen) that are released into the atmosphere that may be harmful to life.

➤ *Primary pollutants*, such as NO_x, SO_x, particulate matter, and CO_2, are released directly from a contaminating source (i.e., a coal-burning power plant or factory).

➤ *Secondary pollutants*, like ozone, photochemical smog, and the nitric and sulfuric acids that become acid deposition, are formed by chemical reactions in the atmosphere involving primary pollutants.

2. Noise and Light Pollution

    i. Many people do not realize that noise and light can be considered forms of pollution. Excess levels of either noise and/or light may cause disruption or harm to eco-systems, as well as human health and well-being.

3. Water Pollution

    i. Water pollution includes any pollutant added to water that results in harmful effects to aquatic ecosystems or water purity (for drinking and irrigation).

    ➤ *Point source pollutants* are those that can easily be identified. Point source water pollutants include effluent overflows from a wastewater treatment plant or a drainage pipe from a factory.

    ➤ *Non–point source pollutants* usually pose a greater problem because they are dispersed, difficult to locate, and as a result are more challenging to prevent.

    — *Runoff* is the most common example of non–point source water pollution. Types of runoff include nutrient-rich agricultural runoff of fertilizers and/or livestock wastes or urban runoffs containing oils and fluids used on and in cars and trucks.

4. Solid Waste

    i. Most of the things thrown away are considered solid waste. Categories of solid waste include paper, metals, glass, plastics, and organic materials.

    ii. Much of the solid waste we produce is *recyclable*, mean-ing it can be collected, melted down, and remade into something new.

5. Toxicity

    i. *Toxicity* is a measure of the harmful effect a substance has on living organisms.

    ii. Toxicity of a substance is determined by a dose/response relationship. This means that, as an organism is exposed to greater amounts or concentrations of a toxin, its harmful effect increases in a predictable manner.

iii. LD-50 (or LC-50 in aquatic systems) is the lethal dose of a toxin that will result in death to 50 percent of a test population.

6. Hazardous Waste

i. Hazardous waste is a category of wastes that includes substances that are highly flammable, poisonous, or corrosive.

➤ Radioactive nuclear wastes are a class of hazardous wastes.

➤ *Carcinogens* are a class of cancer-causing hazardous wastes. Others may cause birth defects (*teratogens*) or genetic mutations (*mutagens*).

➤ All hazardous wastes require special disposal because environmental contamination by these substances leads to dramatic and often catastrophic environmental consequences.

7. Bioremediation

i. *Bioremediation* refers to safely removing certain pollutants or hazardous chemicals from the environment and stored in the tissues of living organisms.

ii. *Phytoremediation* is the most common type of bioremediation. It is the use of hardy, fast-growing plants, like bracken ferns and cottonwood trees, to uptake harmful chemicals like arsenic and mercury from contaminated soils.

VII. Global Change

A. Biodiversity

1. Biodiversity is a measurement of the variety of living organisms in a system. Not only is it a count of the number of different species (species richness), it is also a measurement of the relative abundance of each species (species evenness).

2. Types of biodiversity include species diversity, genetic diversity and ecological diversity.

B. Ozone (O_3)

1. Ozone is either a blessing or a curse depending on where in the atmosphere it is located.

   i. *Stratospheric ozone* (good ozone) is essential to life as we know it. It acts as a sunscreen to block excess UV radiation from penetrating into the troposphere, where it can lead to reduced rates of photosynthesis in plants, as well as elevated risks of skin cancer in people and other animals.

   ii. *Tropospheric ozone* (bad ozone) is a harmful pollutant, a by-product of the combustion of fossil fuels and biomass. It is corrosive, and it may cause respiratory irritation and degrade materials like plastics, metals, and fabrics.

   ➤ Ozone is also a component of photochemical smog.

C. Global Climate Change

1. Global climate change is the collection of changes in global temperatures, climatic patterns, and atmospheric events that has been attributed to human activities over the last few hundred years—especially since the industrial revolution. Most of these changes are thought to result from the increase in atmospheric greenhouse gases from burning fossil fuels.

   i. *Greenhouse gases*, like carbon dioxide, methane, water vapor, nitrous oxide, and chlorofluorocarbons, are gases that trap heat in our atmosphere and cause a warming effect called *the greenhouse effect*.

   ii. The greenhouse effect is essential to life as we know it. Without greenhouse gases, Earth would be a frozen planet.

   iii. An increase in the greenhouse effect may cause global temperatures to increase faster than our natural systems can adapt. This threatens the existence of our ecosystems and ultimately our ability to survive on Earth.

PART II

LIFE ON
Earth

Geology Concepts

I. Reasons for Seasonal and Climate Variation

A. Earth's Tilt

  1. The Earth is tilted at approximately 23.5 degrees on its axis.

B. Earth's Rotation

  1. Every year (365.25 days), the Earth makes one full rotation around the sun.

C. Seasons on Earth

  1. Look at the diagram below. During one part of the Earth's trip around the sun, the tilt causes the southern hemisphere to receive more direct sunlight. During this time, it is summer in the southern hemisphere and winter in the northern hemisphere. As Earth continues to travel around the sun, it eventually reaches a position where the northern hemisphere receives more direct sunlight (imagine the diagram below where the sun and the Earth switch positions—the

sun is on the left and the Earth is on the right but still tilted in the same direction). During this time of the year, it is summer in the northern hemisphere and winter in the southern hemisphere.

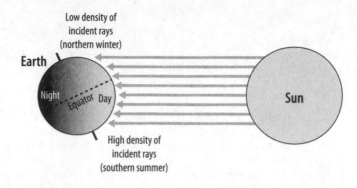

D. Solar Intensity and Latitude

1. Using this same concept, you can see that the equator is always receiving the most direct sunlight and the poles are always getting sunlight at a small angle.

2. Because the rays at small angles tend to spread out over larger areas of the planet, and because the more direct rays tend to be absorbed more directly, the equator is always warmer than the poles.

II. Plate Tectonics and Natural Disasters

A. Layers of the Earth

1. Earth's Core. The Earth's center is a solid core of metals, surrounded by a liquid outer core. The core is composed primarily of nickel (Ni) and iron (Fe).

2. Mantle. The Earth's core is surrounded by the mantle. The mantle is composed mostly of silicon (Si), magnesium (Mg), iron (Fe), and oxygen (O_2).

3. Crust. Directly overlying the mantle is a thin layer called the crust. The crust is made up of basalts and granites.

i. Elements in the Earth's crust by mass:

Element	Percentage of Crust's Mass
Oxygen (O_2)	46
Silicon (Si)	28
Aluminum (Al)	8
Iron (Fe)	5
Calcium (Ca)	4
Sodium (Na)	2

ii. There are two distinct types of crust that make up the Earth's lithosphere: oceanic and continental.

➤ The *oceanic crust* makes up the floor of the oceans' basins; it is both thinner (only 3 to 6 miles) and more dense than the continental crust.

➤ The *continental crust* is the lithosphere underlying Earth's continental landmasses and is both thicker (20 to 30 miles) and less dense than the oceanic crust.

iii. The Earth's crust is broken up into large segments called *tectonic plates*.

➤ These plates are in constant (but very slow) motion and interact with one another in a few predictable ways at their boundaries.

iv. The Earth's plates come together in three major types of boundaries: divergent, convergent, and transform.

➤ *Divergent plate boundaries*, also called spreading centers, are formed when two plates are moving away from one another. This spreading center causes magma from the Earth's interior to rise and cool, making new crust in the form of undersea mountains.

— Earth's most notable divergent boundary is found at the mid-Atlantic ridge running along the seafloor of the Atlantic Ocean between the Americas, Africa, and Europe.

— In general, divergent plate boundaries are associated with seafloor spreading, deep ocean mountain ranges, Africa's rift lakes, and volcanic activity.

➤ *Convergent plate boundaries* are formed when two plates meet and grind together. The plate that is more dense is usually pushed beneath the less dense plate, forming a subduction zone.

— The so-called Ring of Fire is a famous pattern of volcanic and earthquake activity along the continental borders of the Pacific Ocean. This area where the Pacific Plate is subducting beneath surrounding plates accounts for almost 75 percent of Earth's volcanoes.

— The Marianas Trench (Earth's deepest crevice) is part of the Ring of Fire and is an example of a convergent boundary between two oceanic crust plates.

— If both plates are made primarily of less dense continental crust, neither will subduct. Instead, the plates collide and buckle upward into a massive mountain range.

— The most magnificent example of this is the formation of the Himalaya Mountain range as the Indian plate collides with the Eurasian plate.

— In general, convergent plate boundaries are most often associated with deep ocean trenches, volcanoes, and earthquake activity.

➤ *Transform plate boundaries* are formed when two plates, or fractures in a single plate, slide horizontally along one another.

— Often called a transverse fault, the San Andreas Fault, which runs along southern

California and northern Mexico, is one of the most notable transverse boundaries.

— Transform boundaries are most often associated with earthquakes.

v. Intra-plate hot spots are areas of the Earth where there is volcanic activity that cannot be explained by plate boundaries. They are thought to be formed by mantle plumes of rising hot magma beneath one of the plates, which causes softening of the plate and eventually allows magma to break through in the form of a volcano.

➤ The Hawaiian Islands are the most notable examples of intraplate hot spots. Repeated volcanic eruptions cause a buildup of lava rock, which eventually rises above sea level and forms exposed islands.

Test Tip

The only topic from this chapter that regularly appears on the AP Environmental Science exam is that the seasons are caused by the tilt of the Earth on its axis as it rotates around the sun. Be sure to know this. Plate tectonics shows up occasionally, so it's important to know that volcanoes and earthquakes are mostly caused by interactions at plate boundaries. Although plate tectonics has never been used as a topic for a free-response question (FRQ), it would be an excellent topic for a future exam.

Air and Water

I. **The Composition of Earth's Atmosphere**

A. Layers of the Atmosphere

1. The *troposphere* is the layer where we live; most of the atmosphere's oxygen is found here.

   i.   The troposphere extends upward 5 to 9 miles from the Earth's surface at sea level.

   ii. Most weather occurs in the troposphere.

   iii. The troposphere is the most dense layer of the atmosphere (more molecules per unit of volume).

2. The *stratosphere* is directly above the troposphere.

   i.   The stratosphere extends from 9 to 31 miles above the Earth's surface at sea level.

   ii. Stratospheric ozone is vital to trapping and scattering ultraviolet (UV) radiation from the sun (see Chapter 24 for a full discussion of ozone).

   iii. Compared to the troposphere, the stratospheric air is drier and less dense.

   iv. The troposphere and stratosphere together make up the lower atmosphere. Ninety-nine percent of all atmospheric air is found here.

3. The *mesosphere* lies directly above the stratosphere from 31 to 53 miles above the Earth's surface at sea level.

4. The *thermosphere* is directly above the mesosphere, extending from 53 to 372 miles above the Earth's surface.

      i.   Together, the mesosphere and the thermosphere are sometimes called the *ionosphere*.

   5. The *exosphere* is the outermost layer of Earth's atmosphere and acts as the transition zone between Earth's atmosphere and outer space.

      i.   Of all the layers, the exosphere has the lowest atmospheric pressure.

B. Composition of the Atmosphere

   1. The atmosphere is made up of gases. The table below shows the percentage of each gas in the atmosphere.

Atmospheric Gas	Percentage of the Earth's Atmosphere
Nitrogen (N_2)	78
Oxygen (O_2)	21
Water vapor (H_2O)	0–7 (varies by climate)
Carbon dioxide (CO_2)	0.01–0.1 (varies by location)
Ozone (O_3)	0–0.01 (varies by location)

C. Greenhouse Effect

   1. The temperature of Earth's atmosphere is maintained by the absorption of incoming infrared radiation (heat) by atmospheric gases such as CO_2 and CH_4. (See Chapter 25 for a full discussion of the greenhouse effect.)

D. Atmospheric Inversion

   1. An *atmospheric inversion* occurs when the natural temperature gradient of lower atmospheric air is reversed.

      i.   In most cases, the air of the lower atmosphere is warmest near the Earth and becomes cooler as it gets farther from Earth's surface. This is caused by the fact that Earth absorbs infrared radiation (heat) from the sun and radiates it back into the surrounding air.

    ii. When warm air fronts, ocean upwellings, or other atmospheric changes cause the air nearest the Earth to be cooler than the air above, the air becomes more still as normal air currents are suppressed.

> ➤ Fog forms if there is enough humidity for water vapor to condense.

    iii. Inversions may also lead to smog: when dust and particulate pollutants collect in the cooler, still air.

> ➤ A persistent atmospheric inversion in an area with smoke from forest fires can cause the smoke to be trapped, leading to air quality problems for several days.

> ➤ Atmospheric inversions are commonly found in large cities that are surrounded by hills and mountains, like Los Angeles; Mumbai, India; Mexico City, Mexico; and Vancouver, British Columbia.

II. What Causes Weather?

A. Definition

    1. *Weather* is defined as the day-to-day temperature, atmospheric pressure, and precipitation of a given location.

B. Temperature and Density

    1. When thinking about weather, remember that temperature and density are closely related. (See the science review in Chapter 2 for a more detailed discussion of temperature and density.)

    i. Cold air is more dense (more molecules per unit of area). Because of its density, cold air tends to sink through warmer, less dense air. Conversely, warmer air is less dense and rises through cooler air.

    ii. As warm air rises through cooler air, it loses heat and cools. Because cooler air has a lower capacity for holding water vapor, any vapor in the warmer air condenses into water as it rises. Thus, warm air rising tends to return water to the ground in the form of precipitation (rain, sleet, snow, or hail).

iii. As cool air sinks through warm air, it gains heat and warms up. Warm air has a greater capacity for holding water than cool air. As cool air descends toward the ground, it tends to take moisture up and away from the ground as the air warms.

iv. Warm and cold fronts are caused when air masses of differing temperatures interact with one another.

➤ Warm fronts occur when warmer air moves into an area of cooler air. As warm air moves in, it rises and settles over the cool air. As it rises, the warm air cools, often causing drizzling rain because any humidity in the air condenses into liquid water. The warm front tends to add heat to the cooler air beneath it, which increases the temperature and humidity near ground level.

➤ Cold fronts occur when cold air moves in and wedges beneath warm air. The warm air is rapidly lifted as it is replaced by cool air. This causes any humidity in the warm air to condense rapidly and fall as precipitation. For this reason, cold fronts are often associated with a line of rain showers and thunderstorms followed by crisp, cool weather.

v. Tropical cyclones occur when strong thunderstorms combine with extremely moist, warm air (as is often found over tropical oceans). The rapidly rising moist air gives off a great deal of heat as it cools and causes severe downpours.

➤ Cyclones get their name from the fact that as the air rises, it forms a spinning vortex around the area of lowest pressure in the middle of the storm (the eye).

➤ The terms *hurricane* and *typhoon* are regional names for tropical cyclones.

III. Climate

A. Climate

1. *Climate* is defined as the long-term pattern of weather conditions of a particular area.

B. How Earth's Atmospheric Convection Cells Relate to the Biomes

1. In the figure on the next page, look at the equator. As the air is warmed by the sun's radiation, it rises and cools. As the warm, moist air rises and cools, the water vapor it contains condenses and falls as rain. This is why tropical rainforests are common along the equator.

2. At 30° latitude (north and south of the equator), notice that the now cool air falls toward the earth. As that air falls and warms, it has a greater capacity for holding water, and it absorbs water from the Earth's surface, carrying that moisture farther from the equator to 60° latitude. This is why it is common to find deserts at 30° latitude.

3. At 60° latitude, the warm, moist air once again rises and cools, dropping water toward Earth's surface and causing temperate rainforests to be common at this location.

4. The final rotation of the cells results in cool, dry air descending over the poles, warming and pulling water from the land and leading to polar deserts.

5. Combine the idea of the atmospheric convection cells with the fact that the equator is warmer due to more direct infrared (IR) radiation from the sun and with the fact that

the poles are colder due to indirect IR radiation, and you can account for the full array of biomes found on Earth.

The Convection Cell Model of Atmospheric Circulation

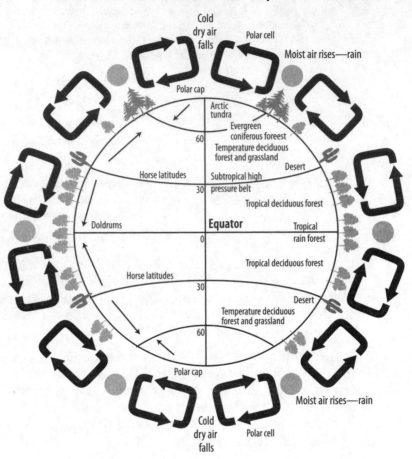

C. Ocean Interactions

1. Convection currents in ocean water work in nearly the same manner as atmospheric convection cells. In general, cool dense water from the poles sinks and moves toward the equator, while warmer, less dense water near the equator rises and floats toward the poles.

2. Because water has an even greater capacity for holding heat than air, these ocean currents carry much of the Earth's heat from place to place.

D. The El Niño Southern Oscillation

1. The El Niño Southern Oscillation (ENSO) is a climate shift found in the tropical equatorial Pacific Ocean.

2. El Niño occurs when the trade winds blowing east to west across the equatorial Pacific Ocean slacken or reverse.

   i. This causes surface waters to be warmer and suppresses nutrient rich upwellings off the northwest coast of South America (near Peru).

   ii. If the upwellings are halted for more than 12 months, populations of plankton, seabirds, fish, and so on, find it difficult to find enough nutrients to survive. As a result, their population numbers decline.

3. ENSO events occur every three to seven years and last from a few months (normal) to as long as 4 years (extreme)!

4. A strong ENSO can affect weather patterns over most of the Earth.

   i. Warmer, dry weather may be found across much of the northern United States, Canada, Brazil, Indonesia, Australia, India, and southeast Africa, leading to:

      ➤ less snow, causing decreased snowmelt in the spring to replenish snow-fed rivers

      ➤ severe droughts

      ➤ increased wildfires

      ➤ increased transmission of diseases due to stagnant warm water and lack of winter kill-off of insect vectors (malaria, dengue fever) and deterioration of fresh water in times of drought (concentration of pollutants, cholera, and other diseases)

   ii. Significantly higher rainfall may occur across much of the southern United States, Cuba, northern Peru, Ecuador, Bolivia, southeast Argentina, and equatorial east Africa, leading to:

➤ suppression of hurricanes in the Caribbean and Atlantic Ocean

➤ flooding which is often followed by landslides

➤ increased transmission of diseases (amoebic dysentery, cholera, giardia) due to contamination of water sources because of flooding and standing water (mosquito breeding).

IV. Global Water

A. Saltwater

1. About 97 percent of Earth's water is considered saltwater.

B. Freshwater

1. Only about 3 percent of Earth's water is considered freshwater. This includes surface water (lakes, rivers, and streams), groundwater, and frozen sources like glaciers, icebergs, and snowcaps.

2. Almost 70 percent of Earth's freshwater exists in one of these frozen sources.

3. Approximately 30 percent of Earth's freshwater exists as groundwater.

4. Less than 1 percent of all Earth's freshwater is surface water (lakes, streams, and rivers).

➤ Lake Baikal in Russia holds 20 percent of Earth's surface freshwater.

C. Agricultural Use of Water

1. Agricultural use is any use of water to grow crops or to maintain livestock. Almost 70 percent of the world's freshwater use is for irrigating crops.

2. Irrigation is most effective when used during the cooler parts of the day (early mornings and evenings). This reduces the amount of water lost to evaporation.

3. There are many methods of irrigation, including passive, spray, and drip irrigation.

   i.   *Passive irrigation* involves digging a system of trenches or canals to divert rain or stream water to crop plants. This works best during times of frequent rainfall, but it is not terribly useful during a dry spell.

   ii. *Spray irrigation* involves using sprinklers or sprayers to water plants. This method is relatively inexpensive to install, is mobile, and can water many plants with relatively few sprinklers. The major drawback is that a large proportion of water is lost to evaporation, especially in hot dry air.

   iii. *Drip irrigation* is often accomplished by using soaker hoses to deliver a slow, steady supply of water directly to the roots of plants. This is especially effective at reducing evaporative water loss, but it is expensive and labor-intensive to install.

4. Aquaculture (raising fish, mollusks, and crustaceans for food, often in large land-based tanks) is another agricultural use of water.

D. Industrial Water Use

   1. Industrial water use is any use of water by a corporation. This includes manufacturing companies (as a solvent), energy companies (as a coolant or source of hydroelectric power), and oil refineries (as part of several chemical reactions).

   2. Approximately 20 percent of worldwide freshwater use is industrial.

E. Domestic Water Use

   1. Domestic water use is any use of water by a private individual, usually within or around the home.

➤ More than half of all water used by an average family is used outdoors. Most of this water goes to irrigation and some is used for washing cars, pets, sidewalks, decks, and driveways.

➤ Flushing toilets accounts for about 10 percent of total domestic water use. This is about one-third of all the water used indoors.

➤ Other indoor domestic water uses include (in descending order of use): laundering clothes; showering; and using faucets, baths, and dishwashers.

2. Water can be conserved in the home by using drought-tolerant plants when landscaping; using washing machines and dishwashers only when the machines are fully loaded; reducing the frequency and length of showers; turning off the faucet when brushing teeth; and installing low-flow toilets, faucets, and showerheads.

F. Groundwater

1. Groundwater accounts for nearly 70 percent of all freshwater.

➤ Groundwater includes water found in the pore spaces between soil and rocks, as well as the flowing water within aquifers below ground surface.

2. *Recharge* is the natural process of water being added back into the aquifer as it slowly *percolates* (moves slowly down through soil and rocks in the ground) into the aquifer.

i. Urbanization can prevent adequate recharge by covering large areas of soil surface with buildings and pavement.

ii. Once ground areas are covered with buildings and pavement, water that would normally be absorbed by the soil and returned to the aquifer instead runs off into gutters, lakes, and streams.

iii. Percolation also helps prevent the buildup of salts around the roots of plants—a common problem caused by excess irrigation—by dissolving and washing them into deeper layers of the subsoil and groundwater.

3. Saltwater intrusion is a common problem in coastal areas. This occurs when saline water enters the aquifer, usually due to the negative pressure established by pumping too much groundwater out for human uses.

G. Global Water Issues

1. Diversion

   i. Water diversion includes the building of dams and canals to redirect, reduce, or halt the flow of a river or stream. This may be used for purposes of irrigation, power generation, or domestic water supply.

   ii. Building dams reduces or stops the flow of water to all points downstream of the dam. Often this results in a large reservoir lake behind the dam. This may lead to several problems in areas downstream of the dam:

   ➤ The natural cycles of flooding and drought are disrupted in these ecosystems.

   ➤ Soils may show reduced fertility because sediments are no longer carried downstream by the river flow.

   ➤ Ecosystems may collapse due to the lack of an adequate water supply.

   ➤ Pollutants and salts that would have been diluted by natural rates of river flow may become concentrated in the smaller amount of water.

   ➤ Shallower water will heat up faster, lowering dissolved oxygen concentrations.

2. Desalination

   i. Desalination is the process of removing the salt from seawater to make freshwater.

   ➤ The Middle East produces and uses about 70 percent of all desalinated water

    ii.   There are two primary methods currently used to desalinate water:

➤ *Distillation.* This ancient and relatively low-tech method simply involves heating the water to the point of evaporation. The evaporated water is then collected and recondensed as freshwater. It is extremely energy-intensive because it takes a great deal of heat to evaporate water.

➤ *Reverse osmosis.* This process involves passing the water through a very fine filtering membrane by raising water pressure on one side of the membrane. Solutes are left behind as pure water passes through. Currently this is an expensive process because plants are expensive to build, they are energy-intensive, and the efficiency is low.

H. Legislation and Conservation

1. The Clean Water Act

    i.   The objective of the Clean Water Act is to restore and maintain the chemical, physical, and biological integrity of the nation's waters. This is accomplished in three ways:

➤ Preventing point and nonpoint pollution sources

➤ Helping publicly-owned water treatment plants to improve wastewater treatment

➤ Working to maintain the integrity of wetlands

2. The Safe Drinking Water Act

    i.   The Safe Drinking Water Act protects the quality of drinking water with two major provisions that:

➤ protect current and potential sources of drinking water from contamination and/or pollution, including surface water and groundwater.

➤ set and enforce minimum standards of purity for tap water provided by wells and municipal water companies.

Frequently, the free-response questions on the AP Environmental Science exam will ask you to discuss a specific piece of legislation. If you are asked to discuss legislation relating to freshwater pollution, or domestic, industrial, or agricultural water-use issues, the Clean Water Act and the Safe Drinking Water Act are your best bets. One of these two could be applied to almost any situation.

REFERENCES

Three Cell Model of Atmospheric Circulation. *http://www.life. umd.edu/classroom/biol106h/L34/L34_climate.html*

Clean Water Act definition from the Environmental Protection Agency. *http://www.epa.gov/agriculture/lcwa.html*

Interactions of Matter
(Soils and Cycles)

 Properties of Soils

A. How Soils Form: Weathering and Erosion

1. Weathering

   i. *Weathering* is the gradual breaking down of rock into smaller and smaller particles.

   ii. *Physical weathering.* Exposure to atmospheric forces causes rock to break down but not to change chemically. For example, freezing water expands within a crack in a rock and causes that crack to enlarge.

   iii. *Chemical weathering.* Chemical reactions change rock to new forms; for example, flowing water dissolves some components of rock, oxidation causes some types of rock to rust, and hydrolysis breaks down silicates to form clay.

   iv. *Biological weathering.* Burrowing animals, plant roots, and so on, force rock to fracture and break down.

2. Erosion

   i. *Erosion* is the movement of soil or rock particles from one place to another. It is usually caused by wind or flowing water.

B. Physical Properties of Soils

1. *Texture* is determined by the particle sizes represented in the mineral component of a soil sample. Particles are classified by size as sand, silt, or clay.

2. *Porosity* is determined by the amount of air space between the particles of soil. Soils with high porosity have larger and/or more air spaces. This allows for development of plant roots and faster infiltration and drainage of water.

3. The moisture/water-holding capacity is the ability of soils to hold water in place.

   i.   Soils with a low water-holding capacity dry out very fast after water is applied.

4. Permeability, or the percolation rate, is essentially the opposite of water-holding capacity. The higher the water-holding capacity, the lower the permeability (water drains through slowly).

   i.   Soils with a low water-holding capacity have a high permeability.

5. Compaction is essentially the opposite of porosity. Soils with fewer and/or smaller air spaces are more compact.

Summary of the Physical Properties of Soil

	Sand	Silt	Clay
Particle size (mm)	2.00–0.05	0.05–0.002	Smaller than 0.002
Porosity	High	Medium	Low
Permeability	High	Medium	Low/impermeable
Water-holding capacity	Low	Medium	High/impermeable
Compaction	Low	Medium	High

C. Chemical Properties of Soils

1. *Organic content* includes leaves, animal wastes, and any materials derived from living (or dead) organisms. Soils with high organic content tend to be more fertile because the decay of organic material returns nutrients to the soil.

2. *Fertility* is a measure of essential nutrients (nitrogen, phosphorus, and potassium) found in a soil sample.

    i.   Nitrogen (N), phosphorus (P), and potassium (K) are the most commonly measured and supplemented soil nutrients, but there are also several micronutrients that are needed in small quantities.

    ii. Organic fertilizers add decayed organic material like composted plants or animal wastes.

> ➤ Decayed organic material increases fertility gradually as the materials decompose.

> ➤ Organic fertilizers also supply the full range of micronutrients and aid in the maintenance of good soil texture.

    iii. Inorganic fertilizers are useful because farmers can target specific soil needs and add only the necessary chemicals.

> ➤ Inorganic fertilizers release nutrients immediately, which can also lead to depletion of micronutrients and soil compaction.

3. *pH* is a measure of the acidity or alkalinity of the soil.

    i.   The normal pH range for growing plants is from 6 (slightly acidic) to 8 (slightly alkaline), although some plants have an optimal pH in either the acidic or alkaline range.

4. *Salinity* is the measure of the salt content of the soil. In general, a good soil has very low salinity. Most plants will not grow well in saline soils.

    i.   *Salinization* is the buildup of salts in the soil and is often caused by fertilizers containing potassium, improper irrigation practices, or infiltration of salt into groundwater.

D. Soil Biota

1. *Soil biota* refers to all organisms living within the soil, including worms, fungi, insects, bacteria, and other microorganisms.

2. In general, biota contribute to soil quality. They provide aeration by tunneling through the soil, and organic content through their wastes and remains.

E. Arability

1. *Arability* is the ability of land to be used for growing crops. To be arable, land needs to have soil that is workable (not compacted) and fertile.

Soil questions frequently appear on each year's exam. Topics most commonly addressed are: soil horizons (diagram and characteristics of each horizon); the fact that tropical rainforests have poor soil for growing crops (and why); types of agricultural practices that lead to soil degradation; and agricultural practices that conserve soil quality.

II. Soil Profiles

A. Soil Horizons

1. Soil layers are divided into *horizons* based on physical and chemical characteristics, as shown in the diagram below.

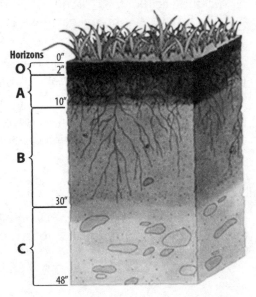

i. O Horizon (Humus)

> ➤ The topmost layer of soil is made up of partially decomposed leaf litter and other plant materials.

> ➤ O Horizon has a very high organic content.

ii. A Horizon (Topsoil)

> ➤ The A Horizon is the most fertile layer of soil.

> ➤ The A Horizon is high in organic content from the decomposition of materials in the O layer.

> ➤ Most biological activity is found in the A Horizon.

iii. B Horizon (Subsoil)

> ➤ The B Horizon layer accumulates iron, clay, aluminum, and other water-soluble compounds that are leached from the O and A horizons as water percolates down through the soil.

> ➤ Deeper plant roots also penetrate into the B Horizon.

iv. C Horizon (Parent Rock)

> ➤ The C Horizon layer contains large unweathered rocks.

> ➤ As rock particles break down, they become integrated into the B Horizon.

B. Soil Profiles by Biome (discussed in order of soil fertility, from high to low)

1. Grasslands

    i.   Grasslands tend to have the most well-developed soils that are rich in organic content with thick O and A horizons.

    ii. Rainfall is low in grasslands (50 to 90 cm/yr) so minerals are not leached out of the topsoil into the water table.

    iii. The decomposition of the grasses during the winter months replenishes the organic content of the soil annually.

  iv. Grassland soil is ideal for growing crops. "America's bread basket" is found in the fertile lands of the North American Prairie.

2. Temperate Deciduous Forests

  i. Temperate deciduous forests accumulate a thick layer of leaf litter in the fall when leaves drop.

  ii. During the winter months, herbaceous plants die and woody plants go dormant.

  iii. At the same time, dead plant material and leaf litter decompose and enrich the soil.

  iv. Due to fairly high rainfall (75 to 150 cm/year), some leaching of nutrients from soil keeps it from being ideal for growing crops.

3. Temperate Evergreen (Conifer) Forests

  i. Temperate evergreen (conifer) forests are found in northern latitudes. These forests have short growing seasons and cold winters.

  ii. The soil is mostly acidic from the breakdown of evergreen needles and has a thick layer of humus in the O horizon. It has fairly high fertility due to low rainfall (30–90 cm/year).

  iii. Temperate evergreen forests make poor farmland because of short growing seasons and acidic soils.

4. Tropical Rainforests

  i. Most people believe that tropical rainforests have very good soils to support the amazing abundance of life found there, but they are wrong!

  ii. Due to the high temperatures and rainfall (200–1,000 cm/year), decomposition occurs very fast.

  iii. Most of the nutrients from organic decomposition are used immediately to support new plant growth, so the majority of nutrients exist above the soil in the rainforest's living organisms.

    iv. The high rainfall also leads to excessive soil leaching and runoff of minerals and nutrients. In areas where rainforests have been cleared for farming, the soils are unsuitable for cultivation.

5. Deserts

    i. Deserts have extremely low rainfall (25 cm or less per year), making conditions poor for soil development and plant growth.

    ii. Most of the rain that does fall evaporates before it infiltrates the soil, leaving behind salts and dissolved minerals. This quick evaporation makes the soil salty, with an alkaline pH.

    iii. Very little organic matter is added to the soil, so fertility is low.

III. Soil Erosion and Desertification

A. Erosion Is a Natural Process

  1. Causes of Erosion

    i. In general, soils that contain more organic material and allow water to infiltrate and drain through tend to have less erosion.

    ii. Agricultural practices leading to loss of organic content, compaction, and reduced plant coverage tend to increase problems with erosion.

  2. Effects of Erosion

    i. Erosion leads to loss of topsoil and reduced arability (desertification).

  3. Prevention and Remediation of Erosion

    i. *Conservation tillage.* Instead of stirring up soil and turning soil over before planting, seeds are inserted into small holes or slits in the ground to reduce disturbance of the soil's O horizon.

ii. *Contour or terrace planting.* Contour (terrace) planting is used when farming on sloped land. Cutting steps or planting on a slope of land reduces the rate of water runoff.

iii. *Agroforestry.* Trees are planted in alternate rows with crop plants, or around the borders of plantings, in order to act as windbreaks and reduce water and wind erosion.

iv. *Soil coverage.* When harvesting crops like corn or wheat, the cut plant material is left to decay on the field. In seasons when a field is not planted, plant-cover crops, like native grasses or nitrogen-fixing legumes, are planted to hold soil in place. Laying down a layer of mulch can also be helpful.

v. *Special irrigation methods.* Methods such as drip irrigation reduce pooling or runoff.

vi. *Less intensive land use.* Instead of planting every field with the same crop each season, crops are rotated from one field to another, and a few fallow (unplanted) fields are left to allow recovery of nutrients and organic matter. If grazing, limit the number of animals per acre to an amount that allows for continued grass coverage (reduce overgrazing).

B. Desertification

1. *Desertification* is the term used to describe the process of land becoming less suitable for cultivation—more desertlike. Areas with low rainfall (such as grasslands) are especially vulnerable.

2. Causes of desertification. Intensive farming techniques requiring high amounts of irrigation and inorganic (commercially produced) fertilizers can cause desertification. Overgrazing on semi-arid grasslands and rangelands is another common cause.

3. Effects of desertification. Loss of organic content, micronutrient depletion (micronutrients are essential elements usually found in small concentrations), and eventually an inability to

grow crops on formerly arable land are some of the long-term effects of desertification.

4. Prevention and remediation of desertification. Many of the techniques used to reduce erosion are also effective at slowing or preventing desertification. The following are also worth noting:

   i. Organic fertilizers. With the use of composted manure or plant materials, micronutrients are added back into depleted soils to improve soil texture, reduce excessive runoff, and prevent rapid infiltration and percolation. Organic material in soil slows water down, so plant roots have a longer time to absorb it.

   ii. Crop rotation. Alternating plants with high nutrient demand and nitrogen-fixing plants, such as peanuts and beans (legumes), can reduce the need for inorganic fertilizers.

   iii. Phytoremediation. *Phytoremediation* is the use of plants to remove chemical contaminants from the soil or to neutralize them. As plants absorb water from the soil, any chemicals dissolved in the water are also taken into the plant, where they accumulate in plant tissues. In some cases, the plant may produce and release chemicals that neutralize chemical contaminants.

   iv. Desalinization. Salt buildup is commonly involved in desertification. One of the most effective ways to desalinate soils is to flood the soil with water and then allow it to slowly infiltrate the soil so that it leaches the salts at the surface to deeper layers and eventually into parent material or the aquifer.

IV. Biogeochemical Cycles

A. The Law of the Conservation of Matter

1. The law of the conservation of matter states that matter may not be created or destroyed. Matter changes forms and

may chemically recombine, but the amount of matter in the universe remains constant.

2. All forms of matter cycle through the environment and take on a variety of forms.

B. Atmospheric Cycles (Cycles with a Gaseous Phase)

1. *The Nitrogen Cycle.* Nitrogen is the most abundant gas in the atmosphere. Nitrogen is often a limiting factor for life in ecosystems because all living organisms need nitrogen to build essential molecules such as nucleic acids (DNA and RNA) and proteins (made up of amino acids). The problem is that atmospheric nitrogen (N_2 gas) is not very chemically active, so it is not a form that can be used by organisms. As a result, a series of chemical transformations is necessary to convert N_2 gas into more usable forms, and then ultimately back into N_2 gas again to continue the cycle.

   i. *Nitrogen fixation.* Soil bacteria and *rhizobium* bacteria living in root nodules of leguminous plants (beans, peas, peanuts, and soy) convert gaseous nitrogen (N_2) from the atmosphere into ammonia (NH_3), which decomposes into ammonium ions (NH_4^+).

   ii. *Ammonification.* Bacteria and fungi in the soil convert nitrogen residues from animal wastes and decomposition into ammonia (NH_3), which decomposes into ammonium ions (NH_4^+).

   iii. *Nitrification.* *Nitrosomonas* bacteria convert ammonium ions (NH_4^+) to nitrite (NO_2^-) then *nitrobacter* bacteria convert nitrite to nitrate (NO_3^-).

   iv. *Assimilation.* Plant roots take up ammonium ions (NH_4^+) and nitrate ions (NO_3^-) for use in making essential molecules.

   v. *Denitrification.* Denitrification is almost the reverse of nitrification. Soil bacteria convert nitrite (NO_2^-) and nitrate (NO_3^-) into nitrous oxide gases (N_2O) and nitrogen gas (N_2), which is returned back into the atmosphere.

2. *The Carbon Cycle.* Like nitrogen, most carbon is found in the gaseous phase in the atmosphere.

   i.   Two major processes of the carbon cycle include photosynthesis and respiration.

> ➤ Photosynthesis. Plants absorb carbon dioxide (CO_2) from the air and use it in the process of photosynthesis, eventually releasing oxygen as a by-product.

> ➤ Respiration. All aerobic organisms (animals, plants, bacteria, and fungi) need oxygen for the process of cellular respiration. CO_2 is released as a by-product of the respiration process.

   ii.   Fossil fuels such as coal and petroleum are remnants of formerly living plants and animals. Both have high concentrations of carbon. Burning fossil fuels releases carbon into the atmosphere.

   iii.   Carbon sinks include forests, oceans, and carbonate deposits. Carbon sinks remove carbon from the atmosphere.

   iv.   Atmospheric carbon (CO_2) is a powerful greenhouse gas and is implicated in global warming.

3. *The Water Cycle (Hydrologic Cycle).* Ninety-seven percent of Earth's water is found in the oceans. This is the major reservoir for water, but water is constantly changing form and moving from one area of the Earth to another. The water cycle is primarily driven by the process of evaporation.

   i.   *Evaporation* is when liquid water changes into water vapor (gas). Heat from the sun is the major source of energy for evaporation. About 90 percent of all the water vapor in the atmosphere comes from the evaporation from surface water (oceans and freshwater) and land.

   ii.   *Transpiration* occurs when water enters the atmosphere from plant leaves. This accounts for the remaining 10 percent of atmospheric water vapor.

iii. *Condensation* occurs when atmospheric water vapor changes to liquid form. This process is how clouds and fog form. Both are made up of tiny droplets of water suspended in the air. Condensation also leads to precipitation.

iv. *Precipitation* is any type of falling H_2O. Types of precipitation include rain (liquid water), sleet (freezing rain), snow (partially frozen water), and hail (balls of ice).

v. Runoff (ocean storage). Much of the water that falls to the Earth lands in the oceans, lakes, rivers, or streams. The water that falls to land will either be absorbed by the land (see the discussion of infiltration in this chapter) or be returned to a body of water as runoff. *Runoff* is when falling water flows across the land (usually pulled by gravity) and eventually makes it back to an ocean, lake, river, or stream. This process is important because many chemicals will dissolve in this water as it flows. Thus, a great deal of water pollution comes from runoff originating in agricultural or industrial areas (see Chapter 22).

vi. *Infiltration and percolation.*

➤ *Infiltration* is the process of water being absorbed into the ground. This process is essential for plant roots to uptake water.

➤ *Percolation* is the downward movement of infiltrated water through the layers of soil, subsoil, and bedrock to the aquifer.

vii. Developing land for human uses (urbanization) affects the water cycle:

➤ Infiltration and percolation are reduced because buildings and pavement cover land, so water runs off into local waterways. As a result, flooding after heavy rains is often a problem.

➤ Runoff from agriculture, roadways, neighborhoods, and so on, may contaminate infiltrated water with pollutants like nitrates, phosphates, and toxic chemicals.

➤ Removal of plant life leads to reduced transpiration, which contributes to a drier, often warmer microclimate in urban areas.

4. *The Sulfur Cycle.* Sulfur is an important element in many proteins. It is the chemical responsible for the smell of rotten eggs. The important aspect of this cycle is that sulfur is a pollutant in some of its chemical forms.

   i. Sulfur dioxide is naturally released from volcanic eruptions and forest fires.

   ii. Sulfur is artificially produced and released as a by-product of burning fossil fuels, mining, and the processing of some metal ores.

   iii. Sulfur in the atmosphere may combine with water to make sulfuric acid (which in turn becomes acid rain).

5. *Sedimentary Cycle.* A sedimentary cycle has no gaseous phase.

   i. *The Phosphorus Cycle.* This is the slowest of the biogeochemical cycles. Phosphorus is not very chemically active. It cycles through the land, water, and living things, but it has no gaseous phase. The important things to know about the phosphorus cycle are as follows:

      ➤ The phosphorus cycle is a sedimentary cycle, which means that, unlike most of the other cycles, it contains no gaseous phase.

      ➤ Phosphorus is essential to life because phosphate (PO_4^{3-}) is necessary for making DNA, and cell membranes.

      ➤ Most soils contain only a small amount of naturally occurring phosphate, so it is considered to be a limiting factor for plant growth. Phosphate is only slightly soluble in water, so aquatic plants are also limited by the ability to obtain enough phosphate.

      ➤ Within a human scale of time, most phosphorus is being washed from rocks and stream beds into marine sediments (Earth's major reservoir of phosphorus). We mine sediments and land-based phosphorus to make fertilizers that increase plant growth. Runoff of these fertilizers into water may severely disrupt aquatic ecosystems because this excess nutrient often leads to algae blooms. This is one type of cultural eutrophication.

If you look at pictures of all these cycles in a textbook, they will all look fairly similar. Focus instead on what makes each cycle unique or important to life on Earth. Questions on the AP exam require both basic knowledge of the major process of each cycle and an ability to show how those processes relate to sustaining life on Earth. If you have time to learn only one cycle well, choose the nitrogen cycle. This shows up on the AP exam more than any other cycle, followed by the carbon cycle and then the water cycle.

How Life Is Organized

 Biotic Versus Abiotic Factors in the Environment

A. Biotic

1. *Biotic* refers to anything that is or was living, including plants, animals, fungi, bacteria, protists, and so on.

B. Abiotic

1. *Abiotic* refers to all nonliving parts of an ecosystem, including rocks, soil, water, weather (humidity, precipitation, temperature), human-made structures, and so on.

C. Interplay Between Biotic and Abiotic Components

1. Because matter is infinitely recycled, there is a constant interplay as chemicals move from living things (biota) to the abiotic environment, and back into living things. Here are some examples:

   i.   The availability of abiotic soil nutrients such as nitrogen, phosphorus, and potassium often limit the growth of plants within an ecosystem.

   ii. Adding nutrients to the soil or water may greatly increase plant growth. Although this may be good for growing crops, it can be devastating for aquatic ecosystems because it can result in cultural eutrophication and lead to algae blooms, lowered dissolved oxygen, and fish kills.

    iii. Dead trees that remain standing (snags) and fallen trees and branches provide habitats for a host of birds, small mammals, and invertebrate species.

    iv. The presence of an exotic species that consumes the leaf litter within a forest may result in many changes, including lower soil fertility, decreased topsoil production, increased erosion, greater rates of evaporation, habitat loss, and so on, as a result of the loss of decomposing leaf litter.

    v. Exotic species may be more able to adapt to these abiotic changes and gain further competitive advantage over native species.

II. Hierarchy of Organization (from specific to general)

A. The Organism Is the Basic Unit of Life

    1. At a minimum, an organism can be one complete cell or it can be as large and as complex as a blue whale.

B. Population

    1. A group of organisms, all members of the same species, who interact, share resources, and interbreed to share the same gene pool.

    2. A *population* is usually defined as the individuals of one species found in a specific area of study.

C. Species

    1. A group of organisms who share similar physical characteristics (look similar) and behavioral traits (act similarly), and who can interbreed to produce fertile offspring.

    2. The result of interbreeding organisms of two different species generally produces a hybrid that is sterile.

D. Community

1. A *community* is the collection of all living organisms of all species (animal, plant, fungi, bacteria, protist) that live together in the same habitat, interact with each other, and compete for resources. Interactions among species include the following:

   i. *Competition.* In many cases, a common resource is essential to the life of organisms of several different species. Whether it be fresh water, a food source, or space for burrows or nests, there is often a struggle for access to these resources.

       ➤ *Competitive exclusion* is the idea that two species cannot coexist if they are in direct competition for the same resource. One of the species will always be slightly better adapted to exploiting the resource and will "competitively exclude the other," ultimately driving the other species either to extinction or to shift to some other resource or habitat to survive. This concept is known as Gause's law of competitive exclusion.

       ➤ *Exotic species* are those not originally from the place in which they are now found.

           — Exotics may be introduced by humans who intentionally transport them from place to place (such as rabbits in Australia or starlings in the United States).

           — Exotics can also be transported unintentionally as hitchhikers (such as zebra mussels in the ballast water of transoceanic ships).

           — In some cases, self-directed migrations of organisms may lead them to become exotic species. Armadillos migrated from South America via Mexico into the southern United States, and coyotes are considered invasive species in the southeastern United States, where they have become established due to natural range expansion and human imports.

ii. *Commensalism* is a relationship between species in which one benefits and the other is unaffected. Some common types of commensalisms are housing, feeding, and transportation relationships.

➤ *Housing.* Many "air plants" like bromeliads, lichens, and orchids grow on the trunks and branches of trees but do not harm or benefit the tree.

➤ *Feeding.* Remoras (sucker fish) attach themselves to the bodies of sharks and other large predatory fish. They feed on scraps that come their way as these fish kill and eat their prey. The sharks seem to be unaffected by the presence of the sucker fish.

— Seagulls are human commensals. They are often found following fishing boats in order to feed as nets are emptied of their catch and entrails are thrown overboard after the fish are cleaned.

iii. *Mutualism* is a relationship between two or more species in which all members benefit directly from the interaction. Examples include the following:

➤ Many species of corals found on tropical reefs have a relationship with a photosynthetic algae living within their bodies. The algae perform photosynthesis and share some of the resulting glucose with the coral in exchange for a safe place to live near the water's surface.

➤ *Pollinators and seed dispersers.* Many flowering plants provide nectar or even nutrient-rich pollen and fruits to insects and small mammal species in exchange for the service of dispersing their seeds or pollen some distance away from the parent plant.

➤ *Protective mutualism.* Some plants provide lodging and a food source to ants or wasps in exchange for their protective services. The acacia plant, for example, houses a colony of ants who will charge out of its hollow stems to attack any other animal who attempts to feed on the acacia leaves or stems.

iv. *Parasitism* is a relationship in which one species benefits at the expense of its host.

➤ Many parasites actually live on or in the bodies of larger organisms feeding on their blood or body fluids (i.e., plasmodium, tapeworms).

➤ Cowbirds are a type of nest parasite. They sneak in and lay their eggs in the nest of another bird, leaving the eggs to hatch and be raised by an unsuspecting host parent. Cowbird chicks tend to be larger and more aggressive than the host's own chicks and often outcompete them for food and parental attention.

E. Ecosystem

1. An *ecosystem* is the collection of all living organisms (community) and their physical habitats. It contains all of the biotic and abiotic components of a given area.

   i. Habitat versus niche.

   ➤ A *habitat* is the environment where an organism lives. It includes soil conditions, weather/climate, and any other physical and chemical factors present in the environment.

   ➤ A *niche* is the role an organism plays within an ecosystem. For example, an ant's habitat would be an underground colony made up of tunnels that were dug by it and fellow ant colony members. The ant's niche would be acting as prey for bird species and ant lions, decomposers, aerators of soil, contributors to soil nutrients, etc.

   ii. *Keystone species* are one or more species that are particularly important to ensuring the stability of an ecosystem. Examples include:

   ➤ *Predators of herbivores.* Sea otters, for example, keep populations of sea urchins in check. Without this population control, sea urchins would decimate kelp plants, which are the primary producers in the kelp forest aquatic ecosystem off the Pacific coast of North America. Wolves feeding on caribou in the North American grasslands can also be considered a keystone species.

➤ *Mutualists.* Many species have relationships in which each provides some necessary resource or service to the other. Removal of one of these species may lead to extinction of its partner species, ultimately affecting the entire ecosystem.

— *Pollinators.* A plant provides nectar as food for the pollinator; the pollinator moves from plant to plant spreading pollen.

— *Seed dispersers.* Squirrels eat the acorns of oak trees and bury many of them to eat later. Some of the acorns are never recovered by the squirrel and then sprout and grow after having been planted by the squirrels.

➤ *Environmental engineers.* Some species physically manipulate the environment in such a way that the changes they introduce make the ecosystem possible.

— Beavers are the ultimate environmental engineers. Their dam-building activities alter the flow of river water, creating ponds that provide habitats for a wide array of other species.

— Prairie dogs dig extensive networks of underground tunnels where they live in large colonies. Their tunnels channel rainwater underground, which reduces runoff and soil erosion.

— Alligators dig "gator holes" in the dry winter months. These holes provide refuge ponds for many aquatic species that would otherwise be unable to survive dry conditions.

— Grizzly bears feeding on salmon remove fish from the river to the surrounding land. They leave partially eaten fish and nutrient-rich feces to decompose and build soil fertility.

iii. *Foundational species* are the autotrophs found at the base of an ecosystem's food web. They are usually present in large populations.

➤ Algae and phytoplankton in aquatic systems, and grasses in terrestrial systems are commonly considered foundational species.

iv. *Indicator species*. Some species, by their presence or absence in an ecosystem, can give information about the physical or chemical characteristics of that environment.

➤ *General environmental characteristics*. By looking at what species are present in an area, information about the characteristics of that area may be learned.

— The calamine violet will grow only in soils that have high concentrations of zinc.

— Macroinvertebrates like mayflies, caddisflies and stoneflies are useful indicators of water quality in aquatic ecosystems. The larval stages of these insects are able to survive only in water with high dissolved oxygen, which usually correlates with high water quality.

➤ *Pollution*. Some species are very sensitive to the presence of pollutants.

— Many species of frogs have recently been showing a dramatic increase in physical defects, such as extra or missing legs. These defects have been linked, in some cases, to pollutants in the water the frogs inhabit. Amphibians in general tend to be sensitive to changes in water quality. Their thin permeable skin allows many pollutant chemicals from the water to enter their bodies.

— Lichens are an example of air quality indicator species. There are three classes of lichen:

crustose, foliose, and fruticose (named by their appearance: crusty, leafy, or bearing fruits). Crustose are the most resistant to air pollutants, foliose are more sensitive to air pollutants, and fruticose are the most sensitive. By looking at the distributions of these three types of lichen, some basic information about how clean or polluted the air is in a location can be determined.

➤ *Environmental change.* Some species are more sensitive to changes in the environment. For example, most of the coral species that populate tropical coral reefs can tolerate water within a very narrow temperature range. Many scientists are closely monitoring coral reef health as a possible indicator of global warming.

v. *Exotic/invasive species.* Some species can move into a new territory and beat out the native species in the competition for resources.

➤ When organisms are moved from their endemic or native habitat to a new one, they may establish a population if conditions are within their range of tolerance. In the new habitat, these newcomers are called an *exotic species.*

➤ In some cases, there will be no natural predators, diseases, or competitors to keep the exotic population in check. Exotic species that adapt very well to their new environment are referred to as *invasive* if they alter their newly adopted ecosystem.

➤ Invasive species are usually able to reproduce quickly and thus leave large numbers of offspring, are very tolerant of environmental changes, and tend to be generalist species able to survive off a wide variety of resources.

➤ Examples include plants like Kudzu and Brazillian pepper and animals like the German cockroach, European starling, cane toads, and even feral cats.

F. Biomes*

1. Earth is made up of several types of regional ecosystems that share common animal life and vegetation and are determined by climactic conditions, specifically temperature and precipitation (rainfall and/or snowfall).

2. Earth's biomes are determined by temperature and precipitation (rain, snow, etc.).

➤ As you learned in Chapter 5, the movement of air in the Earth's convection cells combines with proximity to warm or cool ocean currents to determine an area's temperature and amount of annual precipitation.

➤ Areas with high precipitation tend to have forests.

➤ Lower precipitation inhibits the growth of trees. This leads to the development of grasslands, tundras, or deserts.

3. Types of biomes.

i. **Forests.** Forests are areas where woody trees dominate and eventually form a closed canopy and shaded understory.

➤ **Tropical Rainforests.** Rainforests are found mostly within 25° north or south of the equator. Here are some of their defining characteristics:

— Very high rainfall, with two seasons, wet and dry, due to yearlong warm weather.

— Plants tend to be broad-leaved and evergreen; although rainforests contain Earth's highest biodiversity, their soils are nutrient-poor and unsuitable for farmland if the forest is cleared.

— Notable species include parrots and a wide variety of other brightly colored birds, large snakes, monkey and lemur species, orchids, and figs.

---

* **Note:** The following list does not include every possible biome, only the major ones that are likely to show up on the AP exam.

— Locations include Central America and South America along the Amazon River basin, primarily in Brazil; in Africa, along the Atlantic West African coast, from Liberia all the way around to Democratic Republic of Congo, Central Africa along the Congo River basin, and the island of Madagascar; in Asia, Indo-Malaysia, the west coast of India, Assam, Southeast Asia, New Guinea; and Queensland, Australia.

➤ **Temperate Deciduous Forest.** Temperate deciduous forests are found within the middle latitudes, between the poles and the tropics. Here are some of their defining characteristics:

— Thirty to sixty inches of rainfall per year, distributed fairly evenly throughout the year.

— Four distinct seasons that are roughly equal in length and primarily determined by changes in temperature: spring, summer, autumn, and winter. They are prone to frost and snow in winter.

— Broad-leaf hardwood trees like oak, maple, hickory; shrubs like rhododendron and azalea; an understory of shade-tolerant herbaceous plants; ground mosses; and lichens.

— Because of autumn leaf fall and decomposing leaf litter, soils tend to be well developed and rich in nutrients.

— Notable species include squirrels, deer, black bear, bobcat, and birds (such as cardinals, chickadees, hawks, and blackbirds).

— Locations include the eastern United States and central Europe, as well as locations in eastern Asia, the southern tip of South America, New Zealand, and southeastern Australia.

➤ **Boreal Forest (Taiga).** The boreal forest is found in a broad belt across the northern latitudes from 50° to 60°. Here are some of their defining characteristics:

— Twelve to thirty-five inches of precipitation per year, mostly in the form of snow and sleet.

— Long, dry winters and short, moderately warm, wet summers with a growing season of about 130 days.

— Because of the short growing season, trees are evergreen, bearing resin-filled, needle-like leaves.

— Tree species include spruce, pine, and fir with very little understory vegetation.

— Soils tend to be thin, acidic, and nutrient-poor.

— Animal species include moose, grizzly bear, hares, weasels, and birds (such as woodpeckers, eagles, and hawks).

— Locations include a broad band across the northern latitudes of North America (especially Canada and Alaska), northern Russia (Siberia), and Scandinavia.

ii. **Grasslands.** These areas, found in the middle latitudes, are characterized by a lack of any type of closed canopy from woody trees.

➤ Most of the vegetation is herbaceous grasses and flowers, with some shrubs and few trees. Vegetation is limited to fast-growing herbaceous plants due to the lack of rainfall (only 10 to 35 inches annually), a short growing season (100 to 175 days), and naturally occurring fires.

➤ Summers are warm and often humid; winters are cold and dry; soils tend to be thick and extremely rich in nutrients. When cleared, grasslands make excellent farmland.

➤ Notable plant and animal species include grasses, sunflowers, goldenrod, coyotes, buffalo, prairie chicken, dung beetles, and crickets.

➤ Grasslands tend to be found in the interiors of continents and are named according to geographic location:

— Prairie, or the Great Plains in North America (central states such as Kansas, Minnesota, Wyoming, Montana)

— Pampas in Central South America, especially Argentina.

— Steppe in Russia, from Ukraine to Siberia, and parts of central Asia.

— Tropical savanna in Central Africa.

iii. **Wetlands** are areas that are at least partially flooded during part of their annual cycle.

➤ Many remain flooded year-round, while others are dry for a portion of the year.

➤ Plant life in wetlands must be adapted to moist conditions and include species such as lilies, cat tails, iris, cypress, and gum trees.

➤ Wetlands have a very high biodiversity of animals, including several species of amphibians and reptiles; ducks and wading birds; and mammals such as beavers, rats, and minks.

➤ Water may be fresh (swamps, bogs, fens, freshwater marshes), salty (mangrove forests, salt marshes), or a mixture of the two (estuaries).

➤ Wetlands are of great importance because they are rich in nutrients and serve as breeding, nesting, and nursery grounds for many ecologically and economically important species of birds, fish, shellfish, and so on.

iv. **Chapparals** are arid shrub lands that are most notable for their tendency to be shaped by fire. Chapparal landscape is often used as the setting for movie westerns and is characterized by flat or rocky plains and mountain slopes. Here are some of their defining characteristics:

> ➤ Very low rainfall. Chapparals are often found in the zones between deserts and grasslands. The climate is often-called *Mediterranean* because it is hot and dry for most of the year.

> ➤ In years of increased rainfall, fires ironically are more common due to the increased plant growth, which dries and acts as kindling for fires once the rainfall stops. There has been much controversy about the effect of fire suppression on chaparrals, with most people believing that it just increases the possibility of more severe fires later as biomass builds up.

> ➤ Plants and animals are adapted for conserving moisture. Reptiles have thick-scaled skin; mammals tend to be nocturnal; plants like cactus, creosote bush, manzanita, and yucca tend to have hard, thick leaves.

> ➤ Locations include California, the Mediterranean coast of Europe, the western cape of South Africa, and southwestern Australia.

v. **Tundras** are the coldest biome. They may be either at far north latitudes (Arctic tundra) or high elevations (alpine tundra). Here are some of its defining characteristics:

> ➤ Year-round cold and dry conditions, low growing vegetation, a short growing season, and low biodiversity.

> ➤ Because the growing season is short (50 to 60 days) and there is low precipitation (less than 10 inches per year), plants are very slow growing, extremely fragile, and slow to recover from damage.

> ➤ Animal species in the Arctic tundra include the arctic fox, ground squirrel, caribou, and ravens; in alpine tundra, mountain goats are prevalent.

vi. **Deserts** are characterized by extremely hot and dry conditions; deserts are found mostly near the Tropics of Cancer and Capricorn (latitide 30° north and 30° south). Here are some of their defining characteristics:

➤ Rainfall in deserts is less than 20 inches per year but may be as low as 1 inch in the driest deserts.

➤ Temperatures tend to be hot for most of the year, getting as high as 190°F in some areas.

➤ Soils tend to be sandy and fast draining, with no leaf litter or topsoil.

➤ Species include plants that are adapted to the extreme heat and drought like cactus, yucca, and agave; animals with scaled skin such as snakes, lizards, and tortoises; and burrowing animals, such as kangaroo rats and tarantulas.

vii. **Aquatic ecosystems** are found in the water and can be either freshwater or saltwater. (*The following list includes only those aquatic systems you are likely to see on the exam.*)

➤ Ponds and lakes are *inland freshwater ecosystems.* Large, well-developed lakes include four zones of life:

— *Littoral.* The shallow waters along the shoreline of a lake; extends from the high water line to a depth of about 15 feet or to the depth of the *euphotic zone* (the point where enough sunlight penetrates for plants to grow). A variety of plant and animal species are found here, including cattails, rushes, crustaceans, turtles, frogs, and snails.

— *Limnetic.* The layer of surface water in the middle of a lake. The limnetic layer encompasses the lake's euphotic zone. The majority of the lake's photosynthesis is undertaken by phytoplankton in this zone.

— *Profundal.* The layer of open water in the middle of a lake directly beneath the limnetic zone. Here, there is too little light penetration for plant life to grow. Because of this, the oxygen levels are low in this zone.

— *Benthic.* The bottom of a lake. In this zone, there is little to no penetrating light and a buildup of nutrient-rich sediments. Organisms living in the benthos include scavengers and detritus feeders such as worms, mollusks, crustaceans, and ground fish.

➤ Estuary. Often found at the mouth of a river where it drains into a sea, an estuary is a partially enclosed body of water containing a mix of salt- and freshwater.

When thinking about the biomes, consider how the local conditions of each make it more or less suitable for humans. For example, tropical rainforests have extremely poor soils for farming because, although there is abundant rainfall, they are evergreen forests with a year-round growing season; all the nutrients are found above ground in the living organisms. Contrast this with grasslands, which have a relatively short growing season, a winter dormant season, and seasonal fires. Much of the nutrients in grasslands systems are locked in the thick layers of decomposing plant material and soil. As a result, grasslands are excellent for farming in terms of soil, but irrigation is necessary because of the low rainfall.

G. Biosphere

The biosphere covers the entire realm of Earth where life is possible, including the lower atmosphere, the upper lithosphere, and all of the hydrosphere.

You Are What You Eat (Literally!)

Photosynthesis and Cellular Respiration

A. Photosynthesis

1. *Photosynthesis* is the process by which green plants use the chlorophyll in their cells along with ultraviolet radiation from the sun to make glucose (food). In the process, plants take in carbon dioxide (CO_2) and release oxygen (O_2) as a waste product.

2. The chemical formula for the photosynthesis reaction is:

$$6CO_2 + 6H_2O \rightarrow C_6H_{12}O_6 + 6O_2$$

3. Photosynthesis by plants (and plant-like protists) is significant because it produces the oxygen that makes all *aerobic life* (life requiring oxygen) on Earth possible.

4. Because it uses carbon dioxide as a reactant, photosynthesis is also responsible for helping to maintain the carbon dioxide balance in the atmosphere (carbon dioxide is an important greenhouse gas).

B. Cellular Respiration

1. *Cellular respiration* is the process used by aerobic organisms to break down carbohydrate molecules and convert the energy in their bonds to ATP (a chemical form of cellular energy). All cells undergo cellular respiration to unlock the energy contained in their food.

II. Productivity in Ecosystems

A. Primary Productivity

1. *Primary productivity* is a measure of the energy available to an ecosystem. Primary productivity takes the following into account:

   i. Gross primary productivity (GPP). The total amount of energy from the sun that plants convert to organic matter.

   ii. Respiration (RESP). How much of this energy plants use for their own growth and metabolism (through the process of cellular respiration).

   iii. Net primary productivity (NPP). How much of this energy is then available to the consumers who get their energy by eating plants.

   iv. Calculating primary productivity.

   Gross primary productivity = net primary productivity + cellular respiration.

2. Lab experiment. A common lab experiment for environmental science classes is to calculate the GPP in an ecosystem by measuring oxygen production by plants undergoing photosynthesis and cellular respiration (NPP), then adding back the amount of oxygen consumed by respiration by measuring oxygen loss in plants kept in the dark (RESP).

3. *Secondary production* is the conversion of energy in plant material into biomass at the next highest trophic level (consumers).

Math Practice

Dr. Lawrence has set up an experiment in her lab to determine the gross primary productivity of her fish-tank ecosystem. She first measures the net primary productivity to be 5,000 kcal/ m² per year. She then measures the respiration by aquatic producers to be 7,000 kcal/m² per year. What would her calculated gross primary productivity be?

(continued)

ANSWER:
To calculate the gross primary productivity of an ecosystem, use this equation:

$GPP = NPP + RESP$

$NPP = 5,000$ kcal/m^2 per year

$RESP = 7,000$ kcal/m^2 per year

$GPP = 5,000$ kcal/m^2 per year $+ 7,000$ kcal/m^2 per year

$\quad = 12,000$ kcal/m^2 per year

III. Feeding Levels

A. Food Chain

1. A food chain is a simplified visual representation of the organisms that feed on one another in an ecosystem. For example:

Grass → Cricket → Shrew → Snake → Hawk

B. Food Webs

1. Because of the complexity of feeding behaviors in a typical ecosystem, a food web is used to represent the many food chain possibilities.

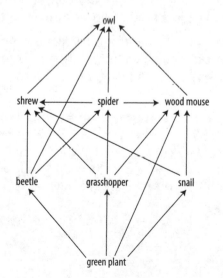

You may be asked to draw a food web from a description of an ecosystem. Remember that, when interpreting or drawing a food web, the arrow is always drawn from food organism to eater, in the direction that energy is moving (from prey to predator). Also remember that you need to draw an arrow from each organism to every other organism that might consume it. Several arrows may be needed from each organism.

C. Trophic Levels

1. *Trophic levels* are feeding levels determined by an organism's position in a food chain. The typical trophic levels in an ecosystem are:

   i. Producers. Organisms that can produce their own food, usually through photosynthesis, form the base of any ecosystem. Grass is an example. These organisms also form the first link in any food chain because they do not eat other organisms, but are usually eaten by other organisms.

   ii. Primary consumer (first-level consumers). These organisms feed on plant material. They are the second link in a food chain. These organisms are herbivores. An example is a cricket.

   iii. Secondary consumer (second-level consumers). These organisms feed on primary consumers and form the third link in a food chain. A shrew is an example.

   iv. Tertiary consumer (third-level consumers). The fourth link in a food chain, these organisms feed on secondary consumers. An example of a tertiary consumer is a snake.

   v. Quaternary consumer. Some food chains go as far as fourth-level consumers, which are organisms that feed on tertiary consumers. They form the fifth link in a food chain. An example is a hawk. Most food chains do not extend beyond this trophic level.

D. Feeding Behaviors of Organisms

1. All organisms are grouped according to their feeding behaviors.

   i.   Autotrophs (Self-Feeders). These are organisms that do not eat. Instead, they use chemical reactions to make the organic compounds needed for respiration. The two types of autotrophs are photoautotrophs and chemoautotrophs.

   ➤ Photoautotrophs. By far the most common autotroph, photoautotrophs include all green plants, green algae, plant-like protists, and many bacteria. They use sunlight as a source of energy to power *photosynthesis* (the conversion of carbon dioxide and water to glucose and oxygen).

   ➤ Chemoautotrophs. These less common organisms (all are bacteria and archaebacteria) use other organic compounds to get their energy. For example, deep ocean vent archaebacteria use hydrogen sulfide gas, whereas soil bacteria, like nitrosomonas and nitrobacter, use nitrates and ammonia.

   ii. Heterotrophs. These organisms must eat other organisms to survive and are grouped according to what types of organisms they eat.

   ➤ Herbivores. Herbivores eat only plant material, including fruits, nectar, leaves, stems, and shoots. Humans who live as herbivores use the term *vegan*.

   ➤ Omnivores. Omnivores eat a mixture of plant and animal materials. Omnivores may eat any plant material, but they also eat animal materials including eggs, and the flesh of other animals.

   ➤ Carnivores. Carnivores eat only animal products. Most predators are carnivores, which feed exclusively on the flesh and eggs of other animals.

➤ Decomposers. This class includes all organisms that feed on dead plant or animal materials. Some classes of decomposers include detrivores (eating dead plant materials) and scavengers (eating dead animal carcasses).

E. Efficiency of Energy Conversions

1. As energy moves through a food chain, from one trophic level to the next, much of the energy is lost in the process. This conversion is often represented as an energy pyramid (E.U. stands for "energy units"):

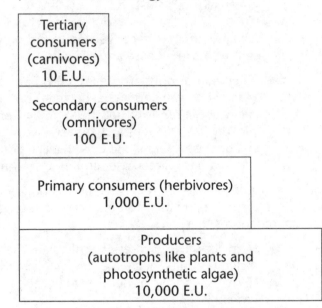

Tertiary
consumers
(carnivores)
10 E.U.

Secondary consumers
(omnivores)
100 E.U.

Primary consumers (herbivores)
1,000 E.U.

Producers
(autotrophs like plants and
photosynthetic algae)
10,000 E.U.

The sun
(the ultimate source of energy in most ecosystems)
1,000,000 E.U.

2. Notice that the efficiency of the conversion of energy from sunlight to producers is approximately 1 percent.

3. The approximate efficiency of the conversion of energy at each step after this (producer to primary consumer and up) is 10 percent.

4. What happens to all the lost energy?

   i.   Some is used by the organism itself for growth and other activities of life.

   ii. Some is lost to the environment in the form of heat.

5. This explains the following:

   i.   It is rare for an ecosystem to have more than four or five trophic levels represented. Beyond that point, there is very little energy left to support life.

   ii. Many animals eat a variety of food types. For example, when you eat a salad, you are acting as a primary consumer at a lower trophic level than when you eat a steak and thus act as a secondary consumer.

   iii. It costs more to produce and buy a pound of meat than a pound of corn.

   iv. Many more people can be supported as vegans than as omnivores using a given area of land.

Math Practice

As part of his graduate research, Mark Waters has measured the amount of energy at the producer level of a grasslands ecosystem to be 720,000 kcal/m² per year. Now he needs to calculate the approximate amount of solar energy coming into this ecosystem. He would also like to calculate the amount of energy available at each trophic level in this ecosystem. He has asked you to show him how to do these calculations. Can you help?

(continued)

ANSWER:

To calculate the solar energy, remember that the energy at the producer level is approximately 1 percent of the incoming solar energy, so do the following:

1. Convert 1 percent to its decimal form = 0.01
2. Use the equation:

   Solar Energy(0.01) = Producer Energy

3. Manipulate the equation to:

   $$\text{Solar Energy} = \frac{\text{Producer Energy}}{0.01}$$

4. Add your numbers into the equation:

   $$\text{Solar Energy} = \frac{720,000 \text{ kcal/m}^2 \text{ per year}}{0.01}$$

   $$= 72,000,000 \text{ kcal/m}^2 \text{ per year}$$

To calculate the energy available at each of the trophic levels above producer, multiply the the value of any trophic level by 10 percent (or 0.1) to determine the approximate energy available at the next level up:

Producers $\times$ 0.1 = Primary Consumers

(720,000 kcal/m^2 per year $\times$ 0.1 = 72,000 kcal/m^2 per year)

Primary Consumers $\times$ 0.1 = Secondary Consumers

(72,000 kcal/m^2 per year $\times$ 0.1 = 7,200 kcal/m^2 per year)

Secondary Consumers $\times$ 0.1 = Tertiary Consumers

(7,200 kcal/m^2 per year $\times$ 0.1 = 720 kcal/m^2 per year)

REFERENCES

Food web diagram from Field Studies Council—Urban Ecosystems. *http://www.field-studies-council.org/urbaneco/urbaneco/introduction/feeding.htm*

The Only Constant Is Change

 I. **How Organisms Change**

A. Diversity

  1. Usually regarded as a measure of the differences among all the organisms in an ecosystem or region, diversity can also be a measure of the differences in ecosystems themselves.

B. Types of Biodiversity

  1. There are different ways to measure the biodiversity of a system. Here are a few:

  i. Species richness. Species richness is a count of the number of different species. For example, if you survey a grassy area and discover a species of mouse, two cricket species, four species of flowers, and three different grass species, species richness would equal 10 species.

  ii. Species evenness. Within each species, how many of each species are represented? For example, a system with 90 of each cricket species, 110 of each species of flowers, 100 of each grass species, and so on, would have much higher species evenness than one in which there were 40 of each cricket, 100 of each flower, and 400 of each grass.

  iii. Genetic variation within a species. The desired state, from a biodiversity standpoint, is to have a high amount of genetic variation within each species.

  ➤ Humans, for example, all look very different and have very different physical abilities. Some people are tall, some people run fast, some people are resistant to malaria. All of these different genetic combinations within the human species make us more adaptable to changes in our environment.

➤ Bottleneck. When a species is reduced to a small number of individuals, the genetic diversity of that species is naturally reduced. If later events allow the species to rebound, there will still be a reduced genetic diversity because all genetic stock can be traced back to the small number of individuals. This leaves the species more vulnerable to extinction due to environmental changes, disease, and so on.

iv. Ecosystem diversity. In a given eco-region, a variety of different ecosystems is usually desired. Ecosystem diversity is reduced by clear-cutting forests to convert the land to farmland and monoculture crop farming.

C. Biodiversity by Biomes/Map Regions

1. In general, areas of highest biodiversity tend to be found near the equator; tropical rainforests, estuaries, and coral reefs all tend to have high diversity. This is likely due to the stable, warm temperatures and relatively abundant supply of water.

2. Conversely, areas with great fluctuations in temperature, or harsh climates that are very cold and/or dry, with low nutrient levels, tend to have much lower diversity. This includes the polar ice caps, the Arctic tundra, and open ocean zones.

The AP exam commonly has a set of questions in which you will have to identify regions on a map. As you study topics like biomes and climate, human population growth, land and water use, and so on, be sure to pull out a world map to see where events are happening. Make sure that you recognize all of the continents and major regions and countries (the United States, China, India, the Middle East, etc.). This knowledge will help you interpret the map questions.

D. Island Biogeography

1. Island biogeography is the scientific theory developed to explain the differences in biodiversity among islands. Essentially it is seen as an interaction between the immigration

of new species to the island and the extinction of species already there. In general the theory states the following:

i. Highest biodiversity corresponds with:

> ➤ large islands due to their tendency to have a wider variety of habitats and resources available to support immigrant species.

> ➤ islands closer to the mainland because it is easier to migrate across a small distance and survive.

ii. Conversely, low biodiversity is expected on smaller islands that are farther away from the mainland.

iii. In addition to explaining actual islands surrounded by water, island biogeography is also useful in understanding habitat islands.

> ➤ Habitat islands are isolated ecosystems—for example, a patch of protected forest surrounded by human development.

> ➤ Island biogeography theory has led to better management of habitat islands like national parks and state forests, including the use of corridors to encourage movement of species from one "island" to another.

E. Biodiversity as an Indicator of Disturbance

1. To achieve the highest biodiversity, just the right amount of disturbance is needed.

> ➤ As we measure biodiversity of a system over a period of time, a decrease is often an indicator of some type of negative disturbance. Pollution, habitat loss, and the introduction of exotic species are all examples of disturbances that may result in lowering the biodiversity of an ecosystem.

> ➤ Ecosystem disturbance is not always bad for biodiversity. The highest levels of biodiversity are often found in ecosystems where disturbances like wildfires or tropical storms occur. This stimulates competition and allows both r-selected and K-selected species to thrive.

F. Evolution

1. *Evolution* may be simply defined as all the changes in a population of a species over time.

2. Natural Selection

    i. First proposed by Charles Darwin, natural selection is the driving force that leads to evolution.

    ii. Individuals that are best adapted to survival and repro-duction in their given environment tend to thrive and produce more offspring. As a result, the *alleles* (genetic information) possessed by these successful individuals become more common in the population over time.

    iii. Natural selection is often stated as survival of the fittest. This works as long as you know that "reproductive fit-ness" is the intent, meaning that an individual leaves lots of offspring. Fitness does *not* have to mean strength, in-telligence, and so on. It could be as simple as being the perfect shade of brown to blend in with the surrounding environment.

3. Adaptation

    i. An *adaptation* is any trait giving an individual a survival or reproductive advantage.

4. Genetic Resistance

    i. *Genetic resistance* is an adaptation in which members of a species become able to tolerate some usually fatal condition. Examples include:

        ➤ Antibiotic resistance. Staphylococcus bacteria becoming increasingly resistant to the antibiotics used to treat staph infections.

        ➤ Pesticide resistance. Pest species developing a genet-ic resistance to the pesticides used to control them.

        ➤ Resistance to disease. For example, people who are heterozygous for the sickle cell anemia trait have a genetic resistance to malaria due to having a pro-portion of misshapen red blood cells. This makes their bodies inhospitable to the reproduction of the plasmodium parasite that causes malaria.

G. Ecosystem Services

1. Ecosystem services include all of the natural processes that contribute to our ability to live on Earth. Some examples include:

   i.   provision of animals and plants for food, clean water, fertile soil, and trees for shelter.

   ii. regulation of Earth's climate, flood control, filtration, and purification of waters.

   iii. supporting human activities through the cycling of matter, decomposition of wastes, and pollination of plants.

   iv. cultural services like providing spiritual and aesthetic support and recreation opportunities.

2. Some human activities that compromise nature's ecosystem services include the introduction of non-native species, pollution, deforestation, and development of land to accommodate a growing population.

II. How Ecosystems Change

A. Climate Change and Species Movement

1. As changes in climate and other abiotic factors occur, species in an area have three options: migrate, adapt, or die.

2. Ecosystems are in a constant state of change as populations of organisms move around, adapt to changing conditions, and disappear altogether.

B. Ecological Succession

1. An example of the changes that take place in ecosystems is ecological succession. This is the process in which a new or disturbed system moves from a state of very little life to a fully developed ecosystem.

   i.   Primary succession. Starting from bare rock, slowly over time, soil is built up and then plants and animals colonize. Drastic disturbances like volcanic eruptions or major mudslides, as well as new island formations, can result in primary succession.

ii. Secondary succession. Disturbances like forest fires, deforestation, and hurricanes may remove most of the organisms in a system but leave behind soil and seeds. The process of nature reclaiming an ecosystem that used to exist but was disturbed is secondary succession.

iii. Pioneer species. Pioneer species are the first species to colonize an area during ecological succession. During primary succession, the most common pioneers are lichens and small ferns and mosses that grow directly on the rocks. During secondary succession, grasses and other fast-growing herbaceous species are often the first pioneers.

iv. Climax community. The end point of succession, a climax community is often a closed canopy, mature forest with tall trees, shade-tolerant undergrowth, and a variety of animal species.

III. How Populations of Organisms Change (Population Biology)

A. Graphing Population Growth

1. Linear Growth. Linear growth is a state where a population grows by a *fixed amount* over time. It may be represented by the sequence: 2, 4, 6, 8, 10, and 12. When shown on a graph, it is a straight, diagonal line.

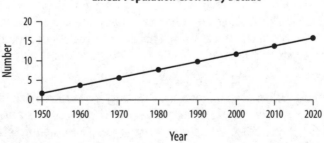

Linear Population Growth By Decade

2. Biotic Potential (Exponential Growth). Biotic potential, or exponential growth, is a state where a population grows by a *fixed percentage* over time. Exponential curves are shaped like the letter J and are characterized by slow growth at first, followed by an explosion of very rapid growth. It may be represented by the sequence 10, 20, 40, 80, 160, 320, 640. This type of growth is characteristic of r-selected species.

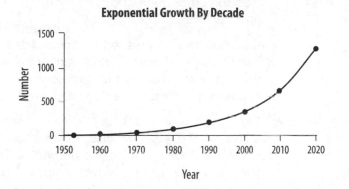

Exponential Growth By Decade

3. Logistic Growth. Most populations show this type of growth. On a graph, this curve is S-shaped. Logistic growth is characterized by initial exponential growth followed by a leveling off of the curve over time. This type of growth is characteristic of K-selected species.

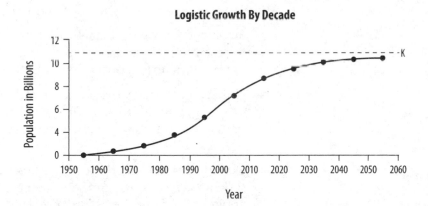

Logistic Growth By Decade

i. Carrying capacity. Carrying capacity is defined as the maximum number of individuals that can be supported indefinitely in a population. It is determined by the abundance of resources in a given environment.

➤ As a population continues to grow, eventually the rate of growth will be kept in check by environmental resistance. For example, food shortages lead to competition and starvation, space becomes limited, and so on.

➤ If a population grows too far beyond its carrying capacity (overshoot), the result may be so damaging to the environment that the overall carrying capacity is lowered for future generations.

➤ On a graph, carrying capacity is often shown as a line marked with the letter K (see above) and is usually the point in a logistic growth curve where the population growth slows or stops. On the graph above, the line representing K is drawn just above the 10 billion mark.

ii. r- and K-selected species. A species may be classified by what type of population growth it usually experiences based on the life history of the species.

➤ r-selected species tend to be species with high biotic potential and rapid population growth rates. These species often show the following traits:

— Small body size, short life span, short time to reach sexual maturity, and short gestation period (time from conception to birth).

— Little parental investment in each individual offspring, large numbers of offspring born per litter or nest, and frequent reproductive events.

— Examples of r-selected species include small rodents such as mice and rats, insects such as aphids, and plant species such as dandelions.

➤ K-selected species tend to have a much lower biotic potential with more stable population sizes that hover closely around the carrying capacity (K). Common traits include the following:

— Larger body size, longer life span, usually a year or more to reach sexual maturity, and a longer gestation period (months rather than days or weeks).

— Greater parental investment, with parents often spending a great deal of time and energy protecting and teaching their offspring for a year or more; fewer offspring per reproductive event (often only one or two); breeding less frequently (often on an annual cycle).

— Examples of K-selected species include elephants, humpback whales, and humans.

➤ Not all species fit neatly into one category or the other. For example, green sea turtles lay several hundred eggs in a nesting season, show no parental care to offspring, and have very high hatchling mortality rates like an r-selected species. But they also have a life span of many decades, take about twenty years to reach sexual maturity, usually breed only every second year, and can weigh hundreds of pounds. These are all characteristics common to K-selected species.

B. Survivorship Curves

1. Another way to classify populations is to look at how many individuals survive at each phase of life. There are three survivorship curves.

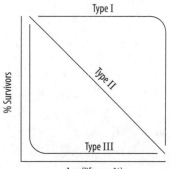

Age (lifespan %)

i. Type I. These species have a high survival rate throughout most of their life span.

> ➤ Most die at an old age.

> ➤ Because there is usually a lot of parental investment in protecting and caring for their young, each offspring has a better chance to survive into adulthood. This is typical of K-selected species.

> ➤ Humans living in more developed countries, zoo animals, and pets also fall into this category.

ii. Type II. These species have a relatively constant death rate throughout their life span.

> ➤ They are equally likely to die at any time in their life.

> ➤ Prey species such as squirrels and small rodents, songbirds, and many reptiles show this type of survivorship.

iii. Type III. These species show a high death rate at the beginning of their life span, which usually levels off once adulthood is reached.

> ➤ Many r-selected species show this pattern of having a lot of offspring with the expectation that very few will survive.

> ➤ Examples include plant species that produce a lot of seeds, oysters, and sea turtles.

PART III

HOW HUMANS USE AND CHANGE Earth

Human Population Dynamics

Human population growth is considered to be the root of all environmental issues. Regardless of whether we are talking about pollution, habitat loss, global warming, or energy resources, the No. 1 thing that we can do to reduce the problem is to slow down the rate of human population growth. More people on the Earth means more pressure on our natural systems. It's that simple!

I. Human Population Dynamics

A. World Population

1. The world's population is almost 7 billion.

2. The current rate of population growth = 1.1 percent.

B. The Three Most Populous Countries in the World

➤ China: 1.3 billion people, growth rate of 0.5 percent

➤ India: 1.2 billion people, growth rate of 1.5 percent

➤ United States: 300 million people, growth rate of 0.97 percent

C. Total Fertility Rate

1. Total fertility rate (TFR) is the number of live births, on average, that each woman of a nation will have during her life.

2. The countries with the lowest TFR also have the lowest population growth.

3. Replacement level fertility equals the TFR necessary to keep population numbers stable. In more developed countries, the replacement level fertility is about 2.1 (taking into account low levels of infant mortality). In countries with higher mortality rates, replacement-level fertility is higher.

4. In general, TFR tends to be lowest in the countries with low infant mortality rates and the highest standards of living.

5. TFR is affected by many cultural and economic conditions. Here are some examples:

    i. Cultures where women achieve a higher level of education tend to see much lower TFRs because women pursue career opportunities after they finish their education.

    ii. Government policies affect TFR, especially if there are regulations regarding birth control availability.

    iii. Many cultures traditionally view a woman who has many children as being more prosperous or "a better wife." This type of tradition has a great deal of power to drive up TFR.

D. Doubling Time

1. When we look at population growth, a common statistic given is *doubling time*. This is the number of years it will take for the population to grow to twice its current size.

2. The rule of 70 is an equation that allows us to calculate doubling time by dividing 70 by the percentage population growth.

$$\text{Doubling time} = \frac{70}{\text{percent population growth}}$$

3. Using the same equation, the percentage population growth can be calculated by dividing 70 by the doubling time.

$$\text{Percentage population growth} = \frac{70}{\text{doubling time}}$$

E. Natural Annual Rate of Increase

1. Natural annual rate of increase is calculated using the difference between the crude birth rate and the crude death rate for a population.

2. Crude birth rate is the number of childbirths per 1,000 people in a year.

3. Crude death rate is the number of deaths per 1,000 people in one year.

F. Growth Rate and Technological Advances

1. Growth rate in the last century has been greatly affected by technological advances.

2. The industrial revolution and the subsequent agricultural and medical advances first led to a population boom as life spans increased and food supply was more stable.

3. In the twenty-first century, people in nations with the greatest access to technology are choosing to have fewer children, resulting in decreasing rates of population growth.

II. Social and Environmental Problems Related to Population Growth and Distribution

A. World Hunger

1. According to most estimates, there is enough food being produced on a global scale to feed all of Earth's population. Why then are people still dying of starvation and malnutrition?

2. Food supply is not distributed equally among all people. In more economically developed countries (MEDCs), people have far more food available than they need. In fact, much goes to waste. In less economically developed countries (LEDCs), there is often a severe shortage of staples such as rice and grain.

3. Poor management of croplands leads to the loss of arable land as soil fertility is diminished.

4. In many LEDCs, population growth has outpaced the people's ability to provide food on a local scale. This is made even worse by the fact that extreme poverty is also common.

B. Disease

1. As population growth leads to more people living closer together (increased density), the opportunities for outbreaks of disease dramatically increase.

2. Some diseases require a vector, a "middle man," that carries the infection from one organism to another. Vector-borne diseases cannot be passed directly from one person to another. Examples include the following:

   i. Malaria is caused by a blood-borne parasite called *Plasmodium*. It is transmitted by the female *Anopheles* mosquito feeding on the blood of an infected person, then passing on the parasite to the next person from whom they feed.

   ii. Bubonic plague: Transmitted from one person to another via fleas.

   iii. Ticks: Common vectors for a variety of diseases, including bacterial infections, chemical toxins, parasites, and viruses. Lyme disease is one notable bacterial disease transmitted though contact with ticks.

3. Some older diseases, thought to be totally eradicated during the twentieth century, are becoming more common due to population growth, complacency about vaccination, and antibiotic resistance. Examples include the following:

   i. Tuberculosis: A bacterial infection of the lungs.

      ➤ Tuberculosis can be spread through the air by tiny particles from the cough or sneeze of an infected person.

      ➤ Another form of tuberculosis is spread through contaminated milk. Pasteurization of milk kills the bacteria before it can cause the disease.

   ii. Malaria: Caused by a vector-borne parasite; common in warm, wet tropical areas near the equator.

      ➤ The best way to control malaria is by eliminating the mosquito vector.

      ➤ DDT was (and still is in many places) commonly used to control mosquitoes, but has been banned in the United States and other countries due to its harmful ecological effects.

   iii. Cholera: A bacterial infection that commonly spreads through unsanitary water supply.

4. Emerging pathogens include newly discovered parasites, bacteria, and virus strains that are causing a host of "new" diseases. These include the following:

   i. HIV/AIDS:

      ➤ First discovered in the 1980s. Although not exactly new, the human immuno-deficiency virus (HIV) is still a recent pathogen in evolutionary terms.

      ➤ A person infected with HIV can be symptom-free for several years before becoming ill with acquired immune deficiency syndrome (AIDS). This makes the virus difficult to eradicate because people may be unaware that they are infected and can unknowingly spread the disease through sexual contact or the sharing of blood.

   ii. H1N1 virus (swine flu):

      ➤ First detected in the United States in 2009 and quickly spread worldwide, this virus caused a great deal of worry especially because of misinformation about how it was transmitted (some people erroneously believed eating pork was the cause).

   iii. SARS (severe acute respiratory syndrome):

      ➤ Caused by a coronavirus and spread through close personal contact, either by inhaling airborne droplets from a cough or sneeze of an infected person, or through touching a contaminated surface and introducing the virus into the body.

➤ In 2003, SARS was first documented in Asia and rapidly became a pandemic, affecting more than 8,000 people throughout North and South America, Europe, and Asia.

5. Currently, the major issue arising in the field of public health is that human populations are no longer geographically isolated. Because of the increase in air travel, disease outbreaks that were formerly contained to a single group of people now have a greater potential for reaching pandemic status.

III. Resource Use and Habitat Destruction

A. Two major aspects of population growth make it an important factor in all other environmental problems: the increasing world population and increasing per-capita resource use.

1. As there are more people on Earth, if per capita resource use remains constant, every additional person puts additional demand on our already limited resources. This is especially true of fresh water and arable land.

2. In reality, per capita resource use is not remaining stable. As more nations continue to reach higher levels of economic development, the people's standard of living is also increasing. So not only are more people being added to Earth's population, each individual is also consuming more and more resources with each passing generation.

IV. More Developed Versus Less Developed Countries

A. Review of Basic Economics Terms

1. Gross national product (GNP). This is the sum value of all the goods and services produced by a nation in one year.

   i. GNP includes all of the companies owned by a nation's citizens, whether they are operating within that country or abroad. For example, a factory producing Ford automobiles in Mexico would still contribute to the GNP of the United States, but not the GNP of Mexico, because Ford is an American-owned company.

2. Gross domestic product (GDP). This is the sum of all goods and services produced within the borders of a nation in one year.

   i.   GDP includes everything produced within a country regardless of who owns the companies involved. For example, the Ford factory in Mexico would contribute to the GDP of Mexico, but not the United States, because the factory is on Mexican soil.

3. *Per capita* means "per person." Calculating per capita GNP or GDP allows the comparison of larger nations to smaller nations in terms of their economic productivity.

   i.   For example, the U.S. GDP is the largest of any nation at just over $15,000,000 billion U.S. However, if we calculate the per capita GDP, the United States barely makes the top ten. (Tiny nations like Qatar and Liechtenstein are more than twice as high!)

B. Standard of Living

1. Standard of living is related to per capita consumption. In general, nations with greater GNP and GDP have an overall higher standard of living for their citizens.

2. In some cases, however, a widening wealth gap between the wealthy and the poor means that a minority of people are extremely wealthy and the majority of people are living at a much lower standard.

C. Age Structure Diagrams

1. Also called population pyramids, age structure diagrams show the number of people at each age group in a population. They tend to be divided with males on one side and females on the other.

Age structure diagrams are one of the most common diagrams to appear on the AP Environmental Science exam. Make sure you know how to interpret these diagrams in order to predict the relative rate of population growth and what level of development corresponds with each type of age structure.

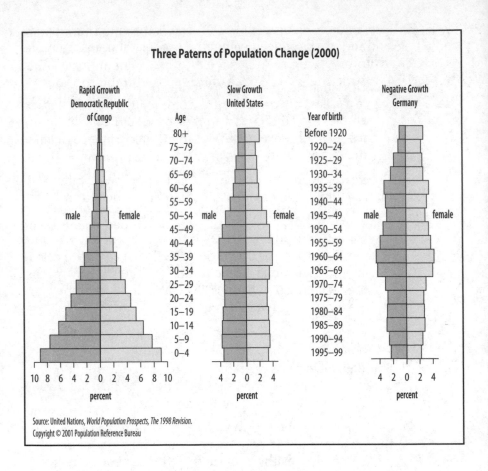

Three Paterns of Population Change (2000)

Source: United Nations, *World Population Prospects, The 1998 Revision.*
Copyright © 2001 Population Reference Bureau

2. Less economically developed countries (LEDCs) tend to have more young people, fewer elderly people, higher birth rates, higher infant mortality, and may also have higher young adult mortality due to political and social conflict (war, genocide).

   i. The age structure diagram of an LEDC shows rapid growth. This means it is very wide at the bottom (indicating a young population), with each succeeding age cohort's bar being significantly shorter than the one below (indicating high mortality rates at all stages of life).

3. More economically developed countries (MEDCs) tend to have aging populations with lower birth rates, low infant mortality, and slower population growth.

    i.   MEDC slow growth. In nations such as the United States and much of Europe, there is a moderate population growth rate of between 0.5 to 1 percent annually. Age structure diagrams for these countries tend to show more balanced numbers among age cohorts, with a slightly wider base and a slightly narrower peak.

    ii.   MEDC zero population growth (ZPG). Some populations find themselves in a state where their population is not growing at all. The age structure diagram for a country like this would show all age cohorts approximately equal (straight sides with no slope). Few countries are currently at ZPG, but several more developed countries are approaching this state.

    iii.   MEDC negative growth. Some nations, such as Italy, Germany, and Greece, are actually losing population over time. An age structure diagram for this situation would be smaller at the bottom (indicating very low birth rate) and larger in the middle and top age cohorts—almost like an inverted pyramid.

4. Although approximately 80 percent of Earth's population resides in LEDCs, those people consume only about 20 percent of resources being used. The remaining 20 percent of Earth's population lives in MEDCs and consumes about 80 percent of resources being used.

D. Demographic Transition

1. Demographic transition is the process in which the population in an LEDC starts to reflect the developmental changes taking place as that country moves toward a state of greater economic development and stability.

2. Most graphs show demographic transition occurring in four stages. Below are two different ways to visualize the changes taking place: by demographic pyramids and a demographic transition birth rate/death rate graph.

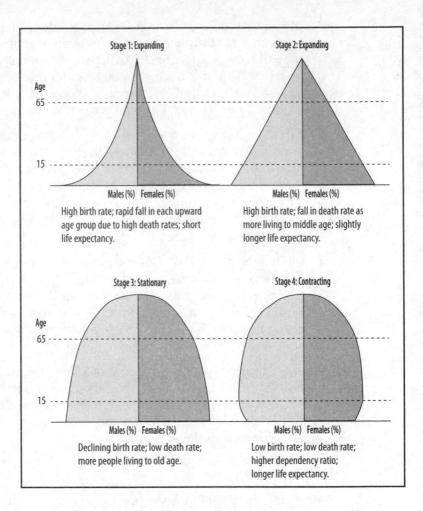

i. The diagram above shows the stages of demographic transition in age structure pyramids.

➤ Stage 1: Shows the typical pyramid for an LEDC, with a high birth rate and high death rate at all age cohorts, and short life expectancy.

➤ Stage 2: As a nation becomes more economically developed, death rates fall. Due in part to better health care, sanitation, and more stable diets, infant mortality decreases and life expectancy begins to increase. The birth rate is still high.

➤ Stage 3: As the nation continues to progress in its economic development, further stability, improved education, and improved medical care lead to a decrease in birth rates. At the same time, more people are living to older ages because of the reproductive time lag caused by longer human life spans; the population continues to grow even as birth rates decrease.

➤ Stage 4: The birth rate continues to decrease and then levels off with higher levels of education and health care availability, especially for women. The population continues to have an older mean age. Eventually this leads to a stabilization of the population to a point of much slower or zero growth. Late in stage 4, after many decades of a stabilized economy and high living standards, countries begin to see declining population sizes.

The graph below shows the four steps of demographic transition in terms of changing birth and death rates and their effect on population growth. This graph is commonly used in AP Environmental Science exam questions.

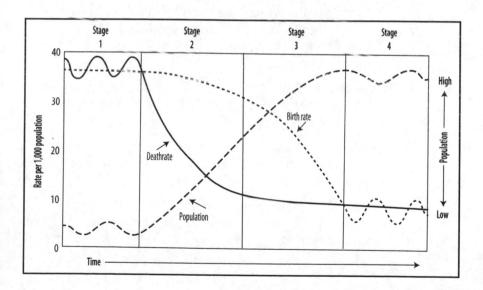

E. Projected Future Growth

1. As we attempt to predict how Earth's human population will look in the future, it is expected that African and Asian countries will have the greatest rates of future growth. China, however, is the exception.

2. Due to the implementation of its one-child policy and strong social pressure toward limiting family size, China's population growth rate has been steadily falling since 1988 and is now well below 1 percent.

F. Strategies for Managing Population Growth

1. India was the first large nation to recognize the need for family planning education and birth control in the 1950s. At that time, India had difficulty successfully implementing its population control programs because of the following factors:

   i. India is a very ethnically and culturally diverse nation, with many languages and ideologies represented, so it is difficult to inform and convince citizens to follow any single national plan.

   ii. Culturally, Indian people tend to value male children, leading many women to continue having children until one or more male children are produced. Many Indian people also value large families, further driving the TFR higher.

   iii. Since the 1980s, however, India's family planning education and birth control campaigns have been much more effective in reducing population growth.

2. Realizing that its population was outgrowing its natural resources, China enacted a strict population control called the one-child policy beginning in 1979.

   i. People who pledge to limit their family to one child receive benefits, such as access to better jobs, education, and even food.

ii. The policy applies only to urban areas. It is accepted that more children may be needed in rural areas to provide labor and support for their families.

iii. The one-child policy has been successful in dramatically decreasing the population growth in China, so much so that it was recently modified to allow a second child in families where both parents were an only child.

iv. The drop in population growth has had some negative effects. People worry that there will be too few young people to fill job vacancies as people retire and that there will be a shortage of caregivers for the elderly. There are also ethical issues involved in governmental control over reproductive rights.

3. Western Europe—Zero Population Growth (ZPG)/ Negative Growth:

i. Some western European nations, such as Greece, Italy, France, and Germany, are finding that their population is either approaching zero growth or, in some cases, declining. While this is good from an environmental standpoint, there are economic concerns associated with ZPG.

ii. In countries with ZPG, the demographic pyramid tends to show an aging population with fewer individuals in the younger cohorts. This may lead to shortages in caregivers for the elderly, and too few qualified workers to fill vacancies caused by retirement.

4. The United States is a world economic leader and, as such, is considered to be an MEDC. Thus, it might be expected that the United States would have zero or even negative population growth like most other MEDCs. However, this is not the case. The United States has a unique set of demographic patterns due to several factors:

i. Among middle- and upper-class Americans there is a trend toward marrying at a later age and having fewer children, which is typical of most MEDCs.

ii. Due to a growing wealth gap, there are also many people living in poverty who lack adequate health care, education, and the financial resources to be considered members of the middle class. For this portion of the U.S. population, women tend to have much higher birth rates; as a result, population growth is much higher.

iii. Unlike many MEDCs, the United States has always had a proportionally large number of immigrants, both legal and illegal. A large proportion of immigrants to the United States come from LEDCs. In most cases, these immigrant groups show similar demographic trends to those seen in their home country, at least for the first generation or two after arriving in the United States. Marrying younger and having larger families drive population growth among many immigrant groups much higher than the U.S. national average.

iv. Taking all these factors into account, the United States tends to have a percentage population growth that is higher than the average MEDC but lower than average for LEDCs.

Math Practice

POPULATION CALCULATIONS

A. Rule of 70:

1. *The United States has a population growth rate of approximately 1 percent. In how many years will the population double if that growth rate remains constant?*

2. *Compare the doubling time of the U.S. population with that of Belize, which has a growth rate of about 2 percent.*

3. *At the current rate of population growth, Earth's population will double in about 64 years. What is the current percentage of the population growth rate?*

B. Growth rate (account for immigration and emigration)

1. *If a population has a crude birth rate of 15 per 1,000 people, and a crude death rate of 3 per 1,000 people, what is the natural annual percentage increase of its population?*

2. *Earth's current population is almost 7 billion and is growing at an annual rate of 1.1 percent. At this rate of growth, how many people would be added in this next year?*

(continued)

ANSWERS:

A. Rule of 70:

1. $\dfrac{70}{1\%}$ = 70 years to double

2. $\dfrac{70}{2\%}$ = 35 years to double. The population of Belize would double twice as fast as the U.S.

3. $\dfrac{70}{64 \text{ years}}$ = 1.1% annual population growth rate

B. Growth Rate

1. Crude birth rate – crude death rate = 15 per 1,000 – 3 per 1,000 = 12 per 1,000. Adjust the decimal to reflect 1.2 per 100, or a 1.2 percent increase.

   ➤ If the crude death rate had been greater than the crude birth rate, the percentage would be a negative number (to reflect that the population would actually be getting smaller).

2. 7 billion = 7×10^9

   1.1% = 0.011

   $7 \times 10^9 \times 0.011 = 0.077 \times 10^9 = 7.7 \times 10^7$

Test Tip

It is almost guaranteed that there will be a growth-rate calculation on the exam. This is good news because they are simple as long as you are familiar with, and remember, the rule of 70!

REFERENCES

Data for population numbers, growth rates, GDP, and GNP from:

U.S. Census Bureau *http://www.census.gov*

The World Factbook. CIA. *https://www.cia.gov/library/publications/the-world-factbook/geos/us.html*

Google Public Data: World Bank, World Development Indicators; last updated December 21, 2010. *http://data.worldbank.org/data-catalog/world-development-indicators?cid=GPD_WDI*

Demographic transition pyramids model from: *http://commons.wikimedia.org/wiki/File:Dtm_pyramids.png*

Demographic transition diagram from the Lewis Historical Society: *http://www.lewishistoricalsociety.com/wiki/tiki-read_article.php?articleId=50*

How We Share Earth's Natural Resources

"The Tragedy of the Commons"

A. Garrett Hardin and the Commons

1. In 1968, Garrett Hardin published an article in the journal *Science* called "The Tragedy of the Commons." His idea was that there is a finite maximum number of people the Earth can support and how we use Earth's resources is an important factor in determining how many people can be sustainably supported.

2. "The commons" Hardin refers to is any resource used by several people (or groups) but owned and controlled by no one.

3. The tragedy is that each person with access to the resource has no reason to limit his or her consumption of the resource, especially if there is no strict regulation with consequences in place for misuse. All it takes is one person or group to take more than his or her share to ruin the sustainability of "the commons" for all.

4. As a result, each user is motivated to get what he or she can while it lasts before others take what's left.

B. Examples of Resources That Are "Commons"

1. International Tuna Fisheries

   i.   Bluefin tuna are extremely valuable fish. Because they migrate thousands of miles and are caught in the open ocean, no single country "owns" the bluefin tuna population. If the United States regulates its catch of the tuna and another country chooses not to regulate its catch, and thus overharvests the tuna, the tuna will eventually become extinct. The United States, however, will have reduced its capacity for profit and allow countries without any regulations to make a fortune by overfishing the tuna population.

2. Overgrazing on Public Land

   i. The U.S. Bureau of Land Management owns huge tracts of prairie land throughout the United States that is used to graze livestock. Much of this land is leased to ranchers for grazing large herds of cattle. Early in the twentieth century, when the land was originally settled, ranchers were encouraged to maximize their profits by keeping more and more cattle on their leased land. As cattle densities increased, the prairie ecosystem was damaged, and this damage led to erosion and desertification.

3. Water Supply

   i. Aquifers (groundwater), large lakes, and rivers are all sources of clean fresh water for drinking, irrigation, and industrial uses. We all depend on these resources, but they can be damaged easily or depleted by any single party who uses them unsustainably.

4. Other Examples

   i. Other examples include atmosphere/air, national forests and national parks, estuaries and wetlands, and the continent of Antarctica (which is managed by a consortium of nations who share land rights).

"The Tragedy of the Commons" shows up at least once on every exam, usually as a multiple-choice question posing a variety of situations and asking which one of the answer choices is an example of tragedy of the commons. Remember to look for the situation of a publicly owned resource, something like the open ocean or the atmosphere, that can't be owned by anyone, is used by many, and is prone to overuse or exploitation.

II. Globalization

A. *Globalization* is the idea that, as people become more and more mobile, and nations depend more and more on each other, the world is becoming a single community.

1. Economics. Many companies based in MEDCs choose to locate factories or manufacturing operations in LEDCs in order to save costs on labor and materials. This makes it possible for companies to offer low prices for their goods.

2. Environmental. The activities of one country often have an impact on many other countries. For example, the United States emits the largest volume of greenhouse gases (notably CO_2 from burning fossil fuels). The resulting increase in global temperatures affects all nations, regardless of their individual contributions to greenhouse gas emissions to the atmosphere.

3. Cultural. Many people see globalization as the mass exporting of Western culture and values around the world. The advent of the Internet has facilitated this type of globalization. Some more traditional cultures, especially those of the Middle East, see globalization as harmful to the traditions and values of their respective cultures.

Feeding Earth's People

I. Agriculture

A. Definition and Most Important Crops

1. *Agriculture* is the production of food by growing plants or raising animals to eat.

2. Corn, wheat, and rice are the most important crops produced to feed the world's population.

B. Types of Irrigation

1. Irrigation is the process of supplying water to crops when rainfall is not enough to provide sufficient moisture for ideal growth of crops. Most farming operations rely on some type of irrigation for at least a portion of their growing season to reduce the effects of weather variations from day to day and year to year.

   i. Spray or Sprinkler Systems

      ➤ Water is pumped through a series of pipes and sprayed out above the ground. One common method is *center broadcast*, in which pipes are suspended over crops and rotated in a circular pattern, spraying water down on the crops as the pipes move around. Best use is on large farms where flexibility of delivery location is important.

      ➤ **Pros:** Many spray systems are fully mobile and can be relocated easily to areas where water is most needed. Often mounted on wheels, these systems can also be motorized and equipped with GPS. These systems are very flexible.

> **Cons:** Evaporation of water is inevitable when water is sprayed through the air to reach crops, especially when you consider that irrigation is usually most needed in warm and dry weather.

ii. Drip Irrigation

> With drip irrigation, water is applied directly to the base or roots of crop plants through perforated pipes or hoses. These pipes are located either on the surface of the ground or buried beneath a shallow layer of mulch or soil. Best use is in areas where water is in short supply.

> **Pros:** Because water is delivered directly to the roots, very little goes to waste, allowing farmers to use less water to achieve the same results. There is almost no evaporative water loss.

> **Cons:** These systems are not mobile. At best, they are laborious to relocate and often time consuming to install.

iii. Flood or Gravity Irrigation

> Large amounts of water are allowed to cover an entire field. Water is often delivered using channels or furrows that are dug among the planted rows of crops. In many cases, water is moved across a sloping field using the force of gravity. Best used to combat soil salinization and/or in areas where water is abundant.

> **Pros:** A lake or stream that can be used as a local water source can be the least expensive method because it requires no specialized equipment or energy to pump water. A large amount of water applied to the soil is also good at dissolving and washing away salts that may accumulate as a result of heavy reliance on irrigation.

> **Cons:** A large amount of water is often wasted as runoff as it washes over the land, and the water may contribute to soil erosion if it is allowed to flow too fast. If water is left standing to slowly

seep in, soil becomes anaerobic because air spaces fill with water, compromising respiration of plant roots and contributing to denitrification of soils.

C. Green Revolution

1. The green revolution introduced new, high-yielding varieties of wheat, corn, and rice into less developed nations like Mexico, India, China, and several others throughout Africa and Asia.

2. The new varieties were selectively bred to be more responsive to irrigation and nitrogen-based inorganic fertilizers. When irrigation and fertilizers were used in combination with the new crop plants, yields dramatically increased and fewer crops failed.

3. The green revolution reached its peak in the 1960s, but it continued through the late 1970s.

4. There is still debate about how successful the green revolution was at reducing hunger. While it undeniably increased crop yields dramatically, there were some problems such as:

   i.   The lack of the necessary infrastructure for properly storing and distributing grains after harvest often led to stored crops spoiling and being wasted instead of feeding the people who were facing food shortages.

   ii. Government corruption led to some nations selling excess food to other nations instead of feeding their own hungry people.

   iii. In some nations, notably India, improper use of the introduced herbicides, pesticides, and fertilizers, as well as their storage containers, led to illness and increased rates of cancer.

   iv. Overuse of intensive agricultural techniques like those of the green revolution is not sustainable in the long term. Inorganic fertilizer use can lead to reduced organic content in soil and compaction, irrigation can lead to soil salinization, pests can develop resistance to pesticides, and so on. Ultimately many of the green revolution projects lasted for a limited time before failing.

D. Better Ways to Feed the People of Earth More Sustainably

1. Vegetarian Diets

   i.   **Pros:** If you remember anything about trophic pyramids and energy moving through ecosystems, you know that eating lower on the trophic pyramid is dramatically more efficient due to the fact that, on average, only about 10 percent of the energy from one trophic level is available to organisms on the next higher level. Therefore, a vegetarian diet or a diet low in meat is more sustainable than a meat-based diet in terms of calories that can be produced per acre.

   ii. **Cons:** The less variety in one's diet, the more prone one is to malnutrition caused by deficiencies of vitamins, minerals, and proteins. Protein deficiency diseases like *kwashiorkor* (causes puffiness and large bloated belly) and *marasmus* (severe loss of muscle tissue) are much more common in areas where diets consist mainly of a single grain staple, like corn or rice, and very little meat.

2. Eating Local

   i.   **Pros:** Eating food that is grown locally is better for both the environment and the local economy. Not only is your money supporting local farmers, it is easier to obtain information about what agricultural techniques were used. Also, less fossil fuels are used to transport the food from origin to point of sale.

   ii. **Cons:** There are fewer options when one is limited by the types of foods that can be produced locally. In the northern climates during winter months, many of the fruits and vegetables that are taken for granted would not be available. In large, densely populated areas such as Chicago and New York City, it would be difficult or impossible to produce enough locally grown/raised food to supply everyone.

3. Genetically Modified Organisms (GMOs)

   i.   **Pros:** GMO crops can often be grown in conditions that would otherwise make it impossible for the crop to grow. Varieties may be adapted to drought, salty soil, or low nutrients. GMOs may be resistant to herbicides, thus allowing for easier elimination of weedy competitors. They may also possess other desirable qualities—like golden rice genetically engineered to contain vitamin A.

ii. **Cons:** Some scientists fear that genetically modified organisms are risky because the crop plants may accidentally be released into natural ecosystems and many of the their modified traits may lead them to be aggressive exotics. Others are concerned about possible negative health effects associated with GMOs. Some simply feel that humans should not tinker with the DNA of organisms to alter their "natural traits" (even though we have been doing this through selective breeding for centuries).

 Pest Control

A. Definition

1. Pest control involves any action humans take to eliminate organisms that are unwanted. Pests may be weed plants that compete with crops, they may be insects and other animals that eat desired crop plants or act as vectors for disease, and so on.

B. Types of Control

1. Chemical Control

i. Pesticide chemicals are named after the target organism. Some examples include herbicides (plant), fungicides (fungi), rodenticides (rodents), and insecticides (insects).

ii. **Pros:** Chemicals are often highly effective at killing the target organisms. Results may be quick and dramatic and may be the only way to control a major infestation of pests.

iii. **Cons:** Chemicals may be toxic to nontarget organisms such as children and family pets. They may also kill natural predators of the target organism. They may also be harmful to the environment, especially if broadcast sprayed.

➤ If overused, target organisms have been known to develop genetic resistance, which leads to the so-called pesticide treadmill. The *pesticide treadmill* is a positive feedback situation where more pesticide is needed to kill the pest. Increased use of the pesticide can lead to development of greater genetic resistance.

➤ DDT is an insecticide that is extremely effective at killing mosquitoes and was widely used starting in the 1940s. Rachel Carson's book *Silent Spring*, published in the 1960s, detailed the harmful effects of DDT on the natural environment (especially birds). The resulting public outcry led to its being banned in the United States in 1972 and banned for agricultural use worldwide soon after. DDT is still used in some places as a deterrent to malaria (mosquitoes are a vector for the parasite that causes the disease).

2. Biological Control

i. Biological control is the introduction of natural enemies to control pest species. These enemies may be predators, parasites/parasitoids, or pathogens.

ii. **Pros:** When it works, biological control is awesome! There is an initial expense of collecting, transporting, and releasing the enemy species. Once they are established, however, very little maintenance is required. Pests are kept in check as they would be in a natural ecosystem without the use of potentially harmful chemicals. Successful examples include the following:

➤ Lady bugs and lacewings released into a garden are voracious predators of mites, aphids, and many other common garden pests.

➤ Dragonflies are predators of mosquitoes. Because they are native to many of the areas where mosquitoes dwell, simply attracting them to the area through environmental modifications (like a source of running water) may be all that is required.

iii. **Cons:** When it doesn't work, biological control of pests can be a disaster. Before attempting a biological control method, extensive research and trials should be done to ensure that the introduced species will not become a pest itself. Some examples of dramatic failures of biological control include the following:

➤ Cane toads in Australia: The life cycle of the toad was not synchronized with the cane beetles it was meant to control. Cane toads are a generalist species, are highly toxic, and have become a threat to many native species. They continue to spread throughout Australia.

➤ Mongooses in Hawaii: Mongooses were introduced into Hawaii to control rodent species. Because most rodents are out at night and mongooses hunt during the day, mongooses have eaten far more native bird species than rats. This is one of the reasons why most of Hawaii's bird species are critically endangered.

3. Other Natural Control Methods

i. Other natural control methods include using organic pheromone control (to disrupt mating), intercropping with plants like chrysanthemums known to repel pests, or using plant-derived sprays like neem oil as pesticides or insect repellants.

ii. **Pros:** When they work, natural control methods are an environmentally friendly alternative to more disruptive or toxic methods. Intercropping in particular has been shown to be effective in many home gardens and organic farms.

iii. **Cons:** Natural control methods may not be as effective or work as fast as inorganic methods. In many cases, use of natural controls requires a deep understanding of the complex natural relationships among species in order to implement them effectively. Timing or placement may be important to determining the outcome.

4. Selective Breeding of New Varieties of Crop Plants

    i. Genetic uniformity increases the likelihood that pesticides will fail and reduces resistance to disease in general. By crossing modern crop varieties with ancestral varieties of the same species, the genetic diversity of the crop species may be increased and result in varieties that are more vigorous.

5. Integrated Pest Management

    i. Integrated pest management (IPM) is the use of a combination of pest management options to increase effectiveness of the regimen while reducing the need to use potentially harmful chemicals.

    ii. Some techniques used for integrated pest management include pest proofing (using mechanical barriers like screens and traps); extensive monitoring for pests to control them before numbers increase; sanitation; changing the physical environment through crop rotation, applying mulch, and so on, to deter pests; and limited use of traditional chemical sprays to those instances when they are needed (and usually in smaller targeted areas and amounts).

III. Rangelands

A. Most of the animals we consume as meat are grazing animals (like beef cattle). They are usually raised on rangeland, where they wander and eat grass for a portion of their lives.

B. More economically developed countries (MEDCs) tend to have diets based heavily on meat.

    1. **Pros:** Meat is high in protein and an excellent dietary source of many essential vitamins (A, B, and D), minerals (iron,

zinc, and selenium), and amino acids needed for proper metabolic functioning. Meat is also a source of fat calories.

2. **Cons:** In most MEDCs, people eat far more meat than they need for a healthy diet. Because meat, especially red meat, tends to be high in saturated fats, it has been linked to many life-shortening conditions such as heart disease and cancer. In terms of ecological efficiency, we have already seen that eating meat is not as efficient as eating a vegetarian diet because of energy loss at each trophic level (see Chapter 8).

   i. Feed lots for cattle and hog/poultry farms. Because of increased demand for inexpensive sources of meat, industrial methods of high-density farming have been developed. There are both environmental and health concerns associated with these techniques. Here are some examples:

   ➤ When raised at high density in less sanitary conditions, animals are more likely to become ill from contagious diseases and infections. To combat this problem, farmers often give their animals maintenance doses of antibiotics even when they are healthy. This is one factor that has contributed to the development of antibiotic-resistant strains of bacteria.

   ➤ To increase growth rates, animals may be fed steroid hormones. Many people are concerned that elevated levels of synthetic hormones in the meat they consume will interfere with their immune systems or disrupt their endocrine systems.

   ➤ High-density farming is often seen as ethically unacceptable. Many question the quality of life of the animals, which are kept in unsanitary conditions and are too closely packed together in their habitats—a far cry from our idealized vision of "life on a farm."

IV. Fisheries

A. Per Capita Fish Consumption

1. Although per capita fish consumption in MEDCs is almost double that of LEDC consumption, fish often makes a more significant dietary contribution in LEDCs: as much as 50 percent of overall protein consumption, especially in island and coastal areas.

B. Regulating How Many Fish Are Caught

1. Overfishing

   i. Overfishing has become a major problem in most areas where people catch and eat fish. As methods of catching fish have become more intense and human population growth has increased, many populations of large, predatory fish are at risk of collapse from overfishing.

   ii. One notable example of overfishing is the bluefin tuna. Prized as a high-quality fish for sushi, each fish brings a premium price. Because these fish migrate across entire oceans, a host of international agreements and treaties have been necessary to regulate the catch limits and thus keep these fish from becoming extinct. Our best efforts so far have fallen short, and the bluefin tuna population may collapse in the near future.

2. Maximum Sustainable Yield (MSY)

   i. When regulating fisheries, the goal is to determine how many fish may be removed from the system each year without harming the ability of a fish population to reproduce and maintain a stable number of individuals. This elusive number is called the *maximum sustainable yield*.

3. Stock Depletion

   i. When people catch more than the MSY, the species population is at risk of becoming reduced in number over time. Limits on time-of-year fisheries are open attempts to allow fish to reproduce undisturbed; size limits also ensure that fish reach reproductive age before being caught. Both of these types of regulations are intended to avoid stock depletion.

C. Fishing Practices

1. Many modern fishing practices have been developed to maximize the amount of fish that can be caught in a limited amount of time. Major drawbacks to these methods include by-catch (organisms such as sea turtles, sharks, and porpoises that are not the target but are killed anyway), environmental degradation, and pollution left when lines or nets break away from the boat and are set adrift. Here are some examples of modern fishing practices and their environmental drawbacks:

i. Trawling. A trawling boat slowly drags a long, funnel-shaped, weighted net behind the boat along the ocean floor, essentially scooping up anything it encounters. Trawling is used to catch shrimp, scallops, cod, and any other fish and shellfish found on or near the ocean floor. This practice severely disrupts the bottom habitat. In the United States, shrimp nets are required to use turtle excluder devices, which eliminate sea turtle by-catch. (All species of sea turtles are protected under the Endangered Species Act.)

ii. Long lines. In this method, fishing lines with lengths of 50 miles or more are reeled out from the back of the boat. Large baited hooks are placed at intervals along the line to catch fish like swordfish, sharks, and tuna. The depth of the line can be adjusted to focus on different types of fish. Cages may be used on the long lines instead of hooks to catch crabs. By-catch or incidental catch is the biggest problem with this method of fishing. Birds, sea turtles, and sharks are often unintentionally caught. A new type of hook, called a circle hook, has been developed to replace the old j-shaped hooks. A circle hook is a good alternative because it greatly reduces by-catch without significantly reducing the target species catch.

    iii. Purse seining. Purse seining is used to catch surface-dwelling fish that swim in large schools. Once a school of fish is spotted, a large circular net is used to enclose the entire school and anything else in the immediate area. Purse seines that are used to catch albacore tuna may also trap schools of dolphins that are swimming with the tuna. This method led to the institution of "dolphin-safe tuna" regulations.

    iv. Drift nets or gill nets. These types of nets are akin to installing a fence in the water to catch anything that swims through an area. These nets are often weighted on the bottom and attached to floating buoys on the top to stay upright.

D. Aquaculture

1. Aquaculture is the practice of raising fish, clams, shrimp, and other aquatic organisms specifically for food or recreation.

2. Aquaculture may take place in large inland tanks or human-made ponds, or it may be practiced by placing large nets or cages in natural bodies of saltwater or freshwater. An increasing percentage of commercially available fish are being raised in aquaculture environments.

3. **Pros:**

    i. All of the problems associated with fishing, such as by-catch and overharvesting, are solved by aquaculture.

    ii. Breeding can be controlled.

    iii. Mercury contamination may be a problem with some large coldwater species of fish. Inland aquaculture operations can raise fish in an environment free of contamination by environmental pollutants.

4. **Cons:**

    i. Inland operations are both labor intensive and energy intensive. As a result, farm-raised fish may be more expensive.

ii. Many of the most desirable fish require huge areas of ocean in order to thrive and breed; as a result, they are not good candidates for aquaculture practices.

iii. Aquaculture operations in open water can cause problems with pollution from wastes and disease due to the high concentration of fish. Farm-raised salmon is a particular problem because parasites from open-water colonies are transferred to wild salmon as they migrate past the nets housing the farmed fish.

Test Tip

Free-response questions that ask about food production topics are very common on the AP Environmental Science exam. In most cases, you will be asked to address environmental or economic (or both) pros and cons of a particular technique. Make sure you read the question carefully and craft an answer that addresses the specific question asked. Whether you are describing environmental or economic effects, be sure to state that clearly in your answer. If you are describing both, state which is which.

Providing for Material Needs:
Forestry and Mining

 I. **Forestry**

A. Forestry is the process of managing forests. The ultimate goal is to maximize human use of trees and forest products without putting the forest in danger of ecological collapse from overuse.

   1. Forests Provide a Host of Ecosystem Services

     i. *Photosynthesis* is the process by which plants take in carbon dioxide (CO_2) gas and, using light from the sun as a source of energy, produce *glucose* (biologically available energy) and oxygen.

     ii. Providing Commercially Valuable Products. Lumber from trees is used for building materials and fuel. Sap provides sweet syrups to eat and latex for glues and rubber. Other products from trees include nuts, such as pecans and cashews, and medicine and cellulose for use in a variety of products, from cosmetics to paper products.

     iii. Habitat for Other Species. Many species, from birds nesting in trees to shade-tolerant understory plants, depend on a closed-canopy forest for habitat.

       ➤ Forest ecosystems tend to have the highest biodiversity of all the biomes.

     iv. Atmospheric and Microclimate Effects. A closed-canopy forest has a distinct microclimate of shade and UV protection and higher humidity due to transpiration. In addition, the trees act as a windbreak.

➤ Large, mature forest trees also act as a sink for atmospheric carbon, not only by converting CO_2 during photosynthesis but also by accumulating it in their tissues, especially the trunks.

v. Improving Soil and Water Quality. The roots of trees act as an anchor for soil, especially in areas with high precipitation. When trees are removed, water carries away topsoil, which in turn leads to erosion, mass movements (landslides), and desertification. By slowing the flow of water across the surface of the soil, trees also facilitate percolation of water through the layers of the soil, increasing groundwater recharge and reducing pollution caused by runoff.

> *The idea of ecosystem services is important and a great topic to bring up when answering a free-response question (FRQ).* **Ecosystem services** *are processes by which natural ecosystems or an environmental process produce resources that humans need. Examples include ocean currents moderating climate, insects pollinating crops, bacteria cycling nutrients through nitrogen fixation and decomposition, and all the examples described in this chapter. There is an increasing movement toward placing a dollar value on ecosystem services as a way to quantify the value of nature.*

2. Logging

i. *Logging* is the process of removing trees from a forest either to clear land for other uses or to use the trees for fuel wood or other commercial purposes.

ii. *Deforestation* is the process of cutting down a significant portion of the trees in a forest and then converting that land to some nonforest use. The most common uses for deforested land are agriculture, grazing livestock, and housing.

➤ *Clear-cutting* is a type of logging in which all of the trees in an area are cut down regardless of size, age, or commercial value. This is the most damaging type of logging and, in most cases, leaves the land in such a condition that it is unlikely to regenerate naturally. Clear-cutting leads to fragmentation of the forest landscape and a decrease in biodiversity.

➤ *Slash and burn* is a type of clear-cutting where trees are cut down and burned on site. It is commonly used on land that will be used for farming because the ash from the burned trees enhances soil fertility.

iii. *Selective logging* is usually considered to be a more sustainable method of cutting trees. Foresters select certain trees to cut while leaving others to continue growing.

➤ Trees may be selected because they are growing too close together (one will be removed to allow the other to grow) or because they have reached a profitable size.

➤ Selective logging still has negative effects for the forest ecosystem because of increased light penetrating to the forest floor. The forest is more able to regenerate if only a few trees are removed at a time.

➤ Another negative impact of selective logging is that roads cut through the forest to transport trees open up greater access to deeper regions of the forest. This often leads to increased human traffic and development along these roads, further eroding the forest and increasing fragmentation.

➤ The most sustainable type of forestry is cutting trees of mixed ages and sizes in small numbers to maintain a natural mixed-age forest with a mostly closed canopy.

3. Classifications of Forest Based on Levels of Disturbance

i. Old-growth forests, also known as primary-growth forests, are forests that have not suffered any major disturbances in hundreds or even thousands of years. They have very large, tall, old trees and very few signs of human interference.

➤ Old-growth forests are characterized as having a closed canopy; a shady, open understory; and the high levels of biodiversity.

➤ Alaska has the largest area of old-growth forests in the United States.

ii. Secondary-growth forests are forests that have regrown after some type of major disturbance. Most of the forests in the United States are secondary-growth forests.

➤ Typical disturbances include human activities (such as logging) and natural events such as storms, insect infestations, and fires.

➤ Forests are classified as secondary growth as long as there continues to be evidence of the disturbance, like dense understory growth and more light penetrating to the ground.

➤ The term *jungle* is often used to describe secondary-growth forests.

4. Forest Fires

i. For most of U.S. history, forest management policy has been to stop any forest fire as quickly as possible. This is known as *fire suppression*. Fire suppression actually increases the risk of extremely hot and fast-burning wildfires because fallen limbs and underbrush accumulate and act as kindling once another fire starts.

ii. Current forest management policy calls for periodic controlled burns, also known as *prescribed burns*. These small burns clear out any debris and brush that have accumulated. This reduces fuel buildup, returns nutrients to the soil, and in many fire-dependent ecosystems, actually stimulates the germination of seeds and maintains the conditions necessary for the ecosystem to thrive.

II. Mining and Extraction

A. Methods of Extraction

1. Depending on the resource being mined, and the depth and type of material around it, several mining techniques are possible. They can be grouped into two categories: surface mining and underground mining.

   i. *Surface mining.* The vast majority of mining is done by removing surface materials, including vegetation, soil, and bedrock, to the depth where the resource is located. Surface mining is commonly used for coal; diamonds; lead; copper; and minerals such as limestone, potash, and rock salt.

      ➤ *Mountaintop removal* is a type of surface mining and is exactly what you would think—the top of a mountain is removed to expose the resource (usually coal) within the mountain.

      ➤ *Open-pit mining* starts on level ground. A large pit is formed out of a series of concentric circles that are carved into the ground by excavating machinery. By far the least expensive and therefore the most common type of mining, open-pit mining is preferred because it is relatively safe if properly planned and executed. It allows large machinery to excavate huge amounts of rock and ore relatively quickly.

      ➤ *Strip mining* is similar to open-pit mining, but long strips (not concentric circles) of land are removed to expose an underlying deposit or "seam" of coal or other resource.

   ii. *Underground mining.* When the desired resource lies deep within the Earth, tunnels are dug into the ground to reach it. This type of mining tends to be less efficient, more dangerous, and more expensive, but if the resource is extremely valuable (such as diamonds or platinum) or in high demand (like coal or base metals such as copper, lead, and iron), it may be worth it.

B. Processing

1. Most of the materials we extract from the Earth are found as ores (rock containing a mixture of valuable minerals or metals and other nonvaluable materials) or impure mixtures in their natural state. Some type of processing is usually required to separate and purify the desired resource.

2. Types of processing include grinding, washing, and chemical treatments.

   i. Grinding. In some cases, ores are ground into smaller particles to separate the individual constituents of the rock.

   ii. Washing. Large amounts of water are used to wash away unwanted particles. This often works because of the differing densities of the individual materials (heavier metals settle and lighter silicates wash away).

   iii. Chemical treatments. In some cases, desired metals may be dissolved and precipitated through a chemical process. These processes usually involve potentially harmful chemicals such as cyanide, mercury, and sulfuric acid.

C. Environmental Regulation of Mining/Extraction

1. Because of the potentially devastating environmental effects of mining, most countries where large-scale mining is done have a host of environmental regulations regarding acceptable practices, recovery of land after a mining operation ceases, and storage and processing of waste products.

2. Overburden Replacement

   i. In surface mining operations, the overburden (consisting of all the rock, soil, and vegetation that were removed to access the resource) must be returned. Mountaintops are rebuilt, and open-pit and strip mines must be refilled and replanted.

   ii. In coal-mining operations, excess sulfur in the soil must be buffered to decrease the acidity of the soil, which would prevent ecosystem recovery. Although this is certainly better than leaving mined areas bare and open, the "recovered" land often takes decades or more to resemble a natural setting.

3. Treatment and Storage of Wastes

    i. Much of the water used in extraction becomes mixed with particles of rock and sediment to form slurry called tailings. If released into waterways, the excess sedimentation and turbidity reduce water quality. *Settling ponds* are used to settle out most of the particles before the water is released.

    ii. Chemicals used in or created by the processes of extraction must be contained and, if possible, chemically altered into less toxic or less harmful products. In some cases, these products are reclaimed so they can be used again.

D. Global Reserves of Strategic Materials

1. Where are all of these resources located? In terms of global reserves of strategic metals, such as rare earth elements, cobalt, chromium, and platinum, the majority of Earth's reserves are being extracted from China, South Africa, and Russia.

E. Recycling Versus Mining

1. For most metals, recycling is the best choice. For example, the amount of energy needed to recycle an aluminum can into a new can is a small fraction of what is required to mine bauxite ore and refine it into usable aluminum. Aluminum, like most metals, is 100% recyclable, with no loss in quality as a result of the process.

REFERENCES

Ecosystem services: A Primer (Action Bioscience). *http://www. actionbioscience.org/environment/esa.html*

Managing Our Public and Private Lands

I. Renewable Versus Nonrenewable Resources

A. Resources are categorized based on whether they can be replenished or used indefinitely within a human span of time (hundreds of years).

  1. *Nonrenewable resources* are used faster than they can be replenished and are expected to run out eventually. Examples include fossil fuels, uranium, diamonds, and metals extracted from the Earth.

      i. Metals such as aluminum (mined as bauxite) and copper can be considered renewable if the ability to recycle and reuse these mined metals indefinitely is taken into account.

  2. *Renewable resources* can be reused or replenished. They include water, soil, forest trees, animals, plants harvested for food, and so on. Land is also considered to be a renewable resource. Renewable resources can be used indefinitely without running out.

      i. However, if the resources are used unsustainably, running out is a distinct possibility for many.

Be sure you understand the difference between renewable and nonrenewable resources. In general, to live more sustainably, we need to shift our consumption from nonrenewable to renewable resources. We also need to stress the importance of managing our renewable resources well so they are not degraded.

II. Managing Public Lands

A. Conservation of Public Lands

1. *Conservation* is the use of a resource in a sustainable manner.

    i. Conservationists feel that we should use our public lands for lumber production, grazing opportunities for livestock, and recreation activities such as hunting and fishing.

    ii. The key is to manage these activities so that they do not lead to the long-term depletion of the resource.

    iii. Many government agencies that oversee our public lands are specifically managed to supply the nation with a sustainable supply of resources.

    iv. Gifford Pinchot, the first head of the U.S. Forest Service, was a conservationist.

2. Bureau of Land Management (BLM)

    i. This agency manages lands primarily in the western United States and Alaska.

    ii. BLM oversees 245 million acres of land and another 700 million acres of below-ground mineral estate.

    iii. BLM is primarily known for managing land for grazing livestock and extraction of minerals, oil, and natural gas.

3. Bureau of Reclamation

    i. Also serving the western United States, the Bureau of Reclamation manages the dams, reservoirs, and water rights associated with the Colorado River and its tributaries. This water is used for irrigation, household use, and recreation.

4. U.S. Forest Service

    i. The U.S. Forest Service manages 193 million acres of forest and grasslands throughout the United States.

    ii. The primary goal of the U.S. Forest Service is to provide a sustainable source of timber.

    iii. Forest Service lands are managed for multiple uses, including extraction.

B. Preservation of Public Lands

  1. *Preservation* is the idea that some land should be set aside and left as wild and untouched as possible.

    i.   The federal government has recognized that, in addition to providing for our material needs, wilderness in its own right is worthy of preservation.

    ii. Lands managed for preservation are usually closed to all consumptive activities like mining, logging, and hunting, although the rules vary.

    iii. John Muir, the founder of the Sierra Club and an early activist for the establishment of national parks, is a notable preservationist.

  2. National Park Service (NPS). In 1872, Yellowstone became the first national park. These lands are set aside to be preserved for future generations.

    i.   National parks are open for recreational activities such as hiking, camping, and fishing. Many national parks do not allow hunting, but some do.

    ii. No logging, mining, or extractive industry is allowed in national parks.

  3. National Wildlife Refuges. Managed by the U.S. Fish and Wildlife Service, wildlife refuges are lands that are set aside specifically to maintain critical habitat for valued species of animals (including fish) and/or plants. Hunting and fishing are often permitted, but the primary activities in wildlife refuges include hiking, photography, and interpretive education.

  4. National Wilderness Preservation Areas. Established as a result of the Wilderness Act in 1964, national wilderness areas are truly managed for preservation. The goal is to maintain a system that is as untouched by humans as possible.

    i.   None of the following activities are allowed on preservation areas: extraction of any natural resources (including wildlife, trees, minerals, fossil fuels, etc.) and construction of roads or permanent structures.

    ii.   Many preservation areas are closed to the public and accessible only to selected scientists or public employees.

III. Wetland Management

A. A *wetland* is an area that generally has standing water covering the land for some portion of its annual cycle. Wetlands include marshes, bogs, and swamps and are characterized by soil that is saturated with water.

B. The Value of Wetlands

1. For much of human history, we have seen wetlands as stinky areas full of snakes, mosquitoes, and alligators. As a result, public policy was primarily focused on draining and filling wetlands so the nutrient-rich soil could be used as cropland. We now know that wetlands provide essential ecosystem services.

    i.   Habitat: Wetlands are among the most biodiverse ecosystems because they offer a wide variety of resources for wildlife. With dense vegetation, water, and abundant nutrients, wetlands are important nursery areas and way stations for migrating animals.

    ii.   Water purification: A wetland acts as a filter for water that passes through it.

       ➤ As water enters a wetland, it slows down, settling out pollutants and nutrient-rich sediments. As water slowly seeps through the soil, impurities are further trapped.

       ➤ Many fast-growing wetland plants also take up toxic chemicals from the environment and store them in their tissues. This process is called *sequestration*.

iii. Flood and erosion control: Wetlands are often found in floodplains and river mouths and act to collect and absorb water as it passes through. As water becomes shallower and more spread out, the velocity decreases. This acts as a protective barrier in times of storm surge or river overflow.

iv. Recreation: The abundance of plants and animals found in wetlands offers excellent recreational opportunities for fishing, bird watching, and nature photography.

C. Human Impact on Wetlands

1. Draining wetlands for development and farming

i. Because they concentrate sediments, drained wetlands leave behind nutrient-rich soil for growing crops. For example, in the early twentieth century, the policy in southern Florida toward the Everglades was to plant water-absorbing melaleuca trees and build a series of drainage canals. Much of this land was then used for growing sugarcane. Water diverted by canals was used to supply the developing urban areas in south Florida.

2. Pollution

i. Excess nutrients like nitrogen and phosphorus used in inorganic fertilizers often lead to cultural eutrophication when they enter wetlands as runoff. The changes in the ecosystem from unchecked eutrophication may trigger additional changes in the wetland, in turn leading to conversion to grasslands or even hammock forests.

3. Protection and Remediation

i. Legislation like the Clean Water Act of 1972 provides protection and guidelines for use and remediation of designated wetlands.

➤ Mitigation banking. When a planned development encroaches on or negatively affects a wetland, one option is *mitigation banking*. This is the process of establishing a trade-off in which one wetland may be damaged or destroyed as long as an equivalent area is restored or protected.

> ➤ Economic incentives and disincentives. Many states have established economic consequences and rewards to influence people's decisions regarding wetland development. Tax breaks for protecting wetlands or donating a privately owned wetland to a preservation group are common. Disincentives like fees and fines for wetland destruction are also effective.

IV. Fragmentation

A. When parcels of land are developed for human use, a common consequence is *fragmentation* of the remaining ecosystem. This creates islands of wilderness broken up by roads, clearings, or developments.

B. Island Biogeography

1. *Island biogeography* (see Chapter 9) is the study of how size and distance from the mainland affects the biodiversity of an island. This theory may also be applied to fragmented terrestrial ecosystems.

2. Smaller fragments that are spread far apart tend to have the lowest biodiversity.

3. To increase biodiversity in fragmented landscapes, a "wildlife corridor" may be created. This is a strip of reforested land that may act as a bridge or migration route between larger fragments.

4. People can develop backyard ecosystems around their homes through planting native plants and providing water features. These backyard ecosystems can also act as a way to help connect corridors.

C. Habitat Islands

1. Many national and state parks are considered to be habitat islands.

V. Legislation to Protect Public Lands

A. National Environmental Policy Act (NEPA) (1969)

1. NEPA requires the filing of an environmental impact statement (EIS) for certain actions, including construction on and development of potentially sensitive lands. The EIS is used as a tool for deciding whether the proposed activity is appropriate.

B. Clean Air and Clean Water Acts

1. Clean Air Act. This act protects land and its inhabitants by regulating the emissions of particulate pollutants and harmful chemicals into the atmosphere. The provisions of the Clean Air Act promote respiratory health; reduce acid precipitation that can be toxic to aquatic and terrestrial ecosystems; and preserve the visibility in cities, parks, and wilderness areas through smog reduction.

2. Clean Water Act. This act regulates water quality through reduction of point and non–point sources of water pollution. Provisions of the Clean Water Act promote ecosystem health by ensuring that water in rivers, lakes, and wetlands is free of toxins and excess nutrients and is maintained at appropriate pH levels.

VI. Urbanization

A. Cities

1. *Cities* are areas where people live in high population density. Cities (or urban areas) are usually characterized by industrial development and the availability of infrastructure and related services (i.e., public water supply, electricity, transportation, healthcare, education, etc.). *Urbanization* is a process where a greater proportion of a population lives in cities.

B. Urban Microclimates/Heat Islands (Why Urban Areas Are Warmer)

1. More asphalt and concrete changes the albedo (reflectivity) of ground surfaces, causing more heat to be absorbed during the day. The heat then radiates back into the atmosphere during the night, keeping nighttime temperatures in cities warmer than in the surrounding natural areas.

2. Urban areas have fewer plants to provide shade and the possible cooling effects of transpiration.

3. More people using air conditioning, cars, and machinery means more heat is generated. Heat is a by-product of combustion.

C. Sustainable Urbanization

1. Although many people consider cities to be harmful to the environment, high-density living can actually be more sustainable than other options. Urban developers may be more likely to adopt more sustainable building codes because their more affluent clients may be more environmentally aware and more willing and able to pay higher costs. Per capita resource use in cities tends to be lower than in suburbia or rural areas.

2. Lower Per Capita Resource Use

   i. People in cities consume fewer gallons of fossil fuels for transportation because they are more likely to use public transportation (buses, subways, trains) or walk to their destination. Compare this to suburban workers, who may drive an hour or more each day (and are usually alone in their cars) just to get to and from work.

   ii. City dwellers use less land per capita for housing. Compare living in a townhouse or apartment to the typical suburban quarter-acre lot with a house and yard. Having no yard means no water used for landscape irrigation, and no pesticides, herbicides, or fertilizers to maintain a lawn. A smaller living space means more efficient heating and cooling. By stacking housing upward in multistory dwellings, the footprint of land per person is also greatly reduced.

3. More sustainable construction. Because urban areas are usually growing, many of our most creative construction alternatives arise in cities.

   i. LEED Certification. The U.S. Green Building Council offers a LEED (Leadership in Energy and Environmental Design) certification program to guide people in the construction of all types of buildings. The LEED program covers energy and water efficiency, sustainable building materials, reductions in indoor pollution, and environmental impact of the building site.

   ii. Cogeneration. *Cogeneration* is the process of recapturing waste heat generated by industrial processes for further uses. For example, waste heat generated by power plants can be redirected toward heating homes.

REFERENCES

National Park Service (NPS)

NPS Units by Type. *http://www.nps.gov/legacy/nomenclature.html*

NPS Overview. *http://www.nps.gov/news/loader. cfm?csModule=security/getfile&PageID=387483*

Bureau of Reclamation: About Us. *http://www.usbr.gov/main/ about/fact.html*

U.S. Forest Service: About Us. *http://www.fs.fed.us/aboutus/*

National Wilderness Preservation System of the United States. *http://www.nationalatlas.gov/mld/wildrnp.html*

Wetlands Definitions/wetlands/U.S. EPA. *http://water.epa.gov/ lawsregs/guidance/wetlands/definitions.cfm*

EPA Wetlands overview *http://water.epa.gov/type/wetlands/out-reach/upload/overview.pdf*

USGBC Intro—What LEED Is. *http://www.usgbc.org/DisplayPage. aspx?CMSPageID=1988*

Energy Basics

I. **The Science of Energy**

A. Comparing Matter and Energy

  1. Matter

    i. *Matter* is defined as any object that occupies space and has mass. Matter is made up of atoms, which may bond together to form molecules. Molecules in turn may combine to form more complex chemicals.

    ii. Earth is essentially a closed system for matter. This means that all the matter on Earth is constantly recycled through the biosphere.

    iii. Because matter is neither created nor destroyed in the universe, there is no significant source of new matter, and nothing that exists ever goes away—it eventually changes into some new form.

  2. Energy

    i. *Energy* is the ability to do work (i.e., apply a force over a distance). Energy is often applied to matter to make some physical or chemical change in that matter.

    ii. Earth is an open system for energy. The sun provides a constant source of concentrated, high-quality energy to Earth.

    iii. Energy does not recycle through the biosphere in the same way that matter does. For example, as energy enters Earth's atmosphere from space, much of the incoming radiation is scattered by the gases in the atmosphere and is reflected back into space. Of the radiation that enters our atmosphere, some warms the air, some is absorbed by water and land surfaces, some reflects off the land and water and radiates back into the air, and some is absorbed by the chlorophyll in green plants for use in photosynthesis.

iv. At every conversion from one form to another, a significant portion of energy is lost to the atmosphere as *waste heat*.

v. The proportion of energy that is used as opposed to that which is degraded or lost is known as *efficiency*.

B. Energy Types

1. Energy exists in two major forms, potential and kinetic.

   i. *Potential energy* is stored energy that can be converted to active or kinetic energy given the right situation.

   ii. The total of all the energy of an object based on its movement (kinetic) or position (potential) is *mechanical energy*.

2. Potential Energy

   i. Energy can be stored for later use in several ways. For our purposes, the three most significant types of potential energy are gravitational, chemical, and nuclear.

   ➤ *Gravitational potential energy* is the energy of an object based on its position and the force of gravity acting on that object. For example, rivers tend to flow from higher elevations to lower elevations. Water is pulled downstream by the force of gravity. If a dam is built to block the flow of a river, the water accumulating behind the dam now has potential energy. As long as the dam holds the water back, the water continues to accumulate gravitational potential energy. If the dam is opened and the water is allowed to flow, the potential energy of the accumulated water is converted to kinetic energy as the water rushes down the river.

➤ *Chemical potential energy*. Much of the energy contained within biological systems, like our bodies, is stored in the chemical bonds of molecules like fats and glucose within our cells. When our cells need to unlock that potential energy, they initiate the process of cellular respiration to convert those molecules into adenosine triphosphate (ATP), which concentrates this energy into larger, more useful packets. As high-energy phosphate groups are removed, from the ATP molecule (to make Adenosine Diphosphate (ADP), then Adenosine Monophosphate (AMP)) energy is released to be used as kinetic energy to power the body's metabolism.

➤ *Nuclear potential energy* is the energy contained within the nucleus of an atom. Some atoms spontaneously split apart in a process called *nuclear fission*. Others can be compelled to nuclear fission through collision with a rapidly moving neutron. When this happens, a great deal of energy is converted from potential to kinetic energy. This process is useful in nuclear reactors for generating electricity.

3. Kinetic Energy

   i. *Kinetic energy* is the energy of motion.

   ii. All moving objects have kinetic energy.

   iii. When the potential energy of matter is put into motion as kinetic energy, work can be done.

C. Forms of Energy

   1. *Thermal energy (heat)*. As matter increases in temperature, the molecules of the atoms making up that matter gain kinetic energy and tend to spread farther apart. This process may result in changes of phase from solid to liquid, to gas, to plasma.

   2. *Electrical energy* is the energy of negatively charged electrons traveling from one place to another through a conduc-

tor. Electrical energy is useful for moving energy across a distance (for example, from a power plant to your home). Once the energy arrives at the desired destination, it can be converted to heat, light, or mechanical energy.

3. *Chemical energy* is the energy stored in the bonds between atoms. For example, the energy locked in the carbon–hydrogen bonds of fossil fuels can be converted to kinetic energy by burning coal or oil to break those bonds.

D. Units of Measuring Energy

1. The *British thermal unit* (BTU) is the amount of heat required to raise one pound of water by one degree Fahrenheit. (This is the most common unit of energy used on the AP Environmental Science exam.)

2. The *joule* is the energy needed to lift an object weighing one Newton by one meter of distance. (The Newton is the metric unit of weight.)

3. A *calorie* is the amount of heat needed to raise the temperature of one gram of water by one degree Celsius. This unit is often confused with food calories (often written as Calorie and called "big calories"), which are actually kilocalories (1,000 calories per 1 Calorie).

4. *Therms* is the unit used to measure the energy in natural gas. One therm is equal to 100,000 BTU and is the approximate amount of energy released by burning 100 cubic feet of natural gas.

5. A *kilowatt hour* (kWh) is the amount of energy used to produce one kilowatt of electricity for a continuous hour. This is the unit most commonly seen on home electricity bills.

Test Tip

It is almost certain you will see one or more energy unit conversion questions on the AP Environmental Science exam. These are common in both the multiple-choice and free-response sections. You do not need to memorize any of the specific conversions because the numbers will always be given in the problem; however, you do need to know how to use them!

Math Practice

E. Thermodynamics

1. *Thermodynamics* is the study of how energy flows through natural systems.

2. First Law of Thermodynamics. The law of conservation of energy states that the amount of energy in the universe is constant. Energy can neither be created nor destroyed. Energy may be transferred from one form to another, but it is never lost and cannot be created anew.

3. Second Law of Thermodynamics. The law of entropy states that, as energy changes from one form to another, it is always moving from a state of more organization to one of less organization.

i.   As energy becomes less organized and concentrated, its usefulness for work is also decreased. (Remember trophic pyramids?) This law is the reason why there are usually no more than five trophic levels in any food web.

ii. This law explains why a perpetual motion machine is impossible. As a machine does work, some of the energy in the system is always lost to friction or heat, so eventually more energy will have to be added from outside the system to allow the machine to continue working.

iii. This law also explains why energy always flows from an object of higher temperature to one of lower temperature.

II.   Energy Conservation

A. Strategies for Conserving Energy

1. Reducing waste in our use/choices. If you closely examine your energy use habits, you will probably find that you are wasting energy all the time. By raising your personal awareness of energy conservation, you can greatly reduce the energy you use without cutting out anything you need.

   i.   Heating and Cooling Your Home

   ➤ Use a programmable thermostat to use less air conditioning during the hours of the day when no one is home.

   ➤ Seal leaky windows, doors, and windows with weather stripping or caulk to avoid air leaks that cause the air conditioning to work more. Add insulation to walls and ceilings and use windows that are double-paned or low-emissive to avoid heat/cooling losses.

   ➤ In designing your home, place windows on south, east or western exposures to allow sunlight to warm the home during the cold winter months.

➤ Pay attention to the exterior color of your home and roofing material. In warm climates, light colors reflect the heat, helping the house stay cooler. In cold climates, darker colors absorb more heat, helping the house stay warm.

ii. Appliances

➤ Even when turned off, many appliances still use electricity as long as they are plugged in. Large appliances like computers, televisions, and stereo systems should be plugged into a power strip that can be turned off when the appliances are not in use.

➤ Even in sleep mode, computers and televisions use energy. Turn off these appliances at night and when you will not be using them for several hours. If you use a computer at school or work, turn it off (including the screen) before you leave for the day.

➤ Turn off lights and other appliances when you are not in the room. Do a quick walk through your house before you leave for the day to ensure that ceiling fans, radios, and lights are not left on.

2. Increase the efficiency of existing technology. In the last few years, the energy efficiency of many of the things we use has been greatly improved. As you purchase new products, compare the energy use of your options before you make your choice. Consider not only the purchase price, but also the long-term cost of using that product every day.

i. Compact Fluorescent Lights (CFLs) and Light-Emitting Diodes (LEDs). When choosing a replacement light bulb, many people still buy incandescent bulbs out of habit. They cost less to purchase (less than a dollar compared to $3 or $5 per CFL). Some people prefer the quality of light produced by incandescent lights. The reality is that CFLs and LEDs are far and away the best lighting option.

➤ Only 10 percent of the energy used by incandescent bulbs produces light. Ninety percent of the energy used by the bulb is used to generate heat! That is why you should never touch an incandescent bulb that has been on recently.

➤ CFLs last, on average, ten times longer per bulb and use 75 percent less energy than incandescent bulbs. The technology in making the CFL bulbs has greatly improved in recent years, putting an end to slow start-up times, washed-out color, and that annoying hum that keep some people from trying CFLs.

➤ LEDs are a newer technology. LED lighting produces a focused beam of light and is excellent for recessed fixtures or task lighting (like a desk lamp). The cost per bulb is much higher than the other two options ($30 or more per bulb). With the overall savings in electricity and extremely long life of the LED fixtures, however, they are arguably the most efficient option in the long term.

ii. Energy Star appliances. *Energy Star* is a rating system that allows consumers to compare the energy use of new appliances to one another. Energy Star-certified appliances meet minimum energy conservation criteria and are good choices when considering the long-term cost of a new appliance.

iii. Gas mileage in cars. With gas prices on the rise, one major consideration when buying a new car is gas mileage. Improvements in lightweight body materials, more efficient engines, and better design mean many cars are getting more than 30 miles per gallon on the highway. Options like hybrids and electric cars offer even greater energy efficiency.

Math Practice

Efficiency problems are simply a matter of working with proportions and percentages. Here are a few practice questions.

1. *If your home uses twenty 100-watt incandescent light bulbs for four hours per day, how many kilowatt-hours of electricity are needed to power the bulbs for one year of use?*

2. *Compact fluorescent light bulbs (CFLs) are 75 percent more efficient than incandescent bulbs. How many kilowatt-hours per year can you save by replacing 20 incandescent bulbs with CFLs?*

3. *A compact hybrid gets 51 miles per gallon of gasoline in highway driving. A luxury sedan averages only 25 miles per gallon on the highway. If gas costs $3.00 per gallon, how much money will a person who drives 10,000 miles per year save by choosing to purchase the hybrid?*

ANSWER:

1. Your answer needs to be in units of kilowatt-hours per year. As you set up your fractions, make sure the units line up to cancel each other, leaving your desired units in the right places!

$$20 \text{ bulbs} \times \frac{100 \text{ watts}}{\text{bulb}} \times \frac{1 \text{ kW}}{1000 \text{ watts}} \times \frac{4 \text{ hours}}{\text{day}} \times \frac{365 \text{ days}}{\text{year}} = \frac{2920 \text{ kWh}}{\text{year}}$$

2. Seventy-five percent more efficient means that only 25 percent of the energy needed to run incandescent bulbs is needed to run CFLs. Simply multiply your answer to Math Practice 1 by 25 percent.

$$\frac{2920 \text{ kWh}}{\text{year}} \times 0.25 = \frac{730 \text{ kWh}}{\text{year}}$$

3. First, note that your answer needs to be in the units of dollars per year. Set up your fractions to cancel units, making sure dollars is on the top of the fraction and year is on the bottom. The rest falls into place.

$$\text{Compact hybrid} = \frac{10,000 \text{ miles}}{\text{year}} \times \frac{1 \text{ gallon}}{51 \text{ miles}} \times \frac{\$3}{\text{gallon}} = \frac{\$588}{\text{year}}$$

$$\text{Luxury sedan} = \frac{10,000 \text{ miles}}{\text{year}} \times \frac{1 \text{ gallon}}{25 \text{ miles}} \times \frac{\$3}{\text{gallon}} = \frac{\$1200}{\text{year}}$$

$$\text{Savings} = \text{cost for sedan - cost for hybrid} = \frac{\$1,200}{\text{year}} - \frac{\$588}{\text{year}} = \frac{\$612}{\text{year}}$$

REFERENCE

Light bulbs (CFL): Energy Star. *http://www.energy-star.gov/index.cfm?fuseaction=find_a_product.showProductGroup&pgw_code=LB*

Nonrenewable
Energy Resources

A. *Nonrenewable energy resources* are those that cannot be replenished. Regardless of how they are used, the supply of the resource will eventually be exhausted. The primary nonrenewable energy resources are fossil fuels and nuclear power.

How Electricity Is Generated in Power Plants

A. Most power plants generate electricity using the same basic process. Some type of fuel is used to produce enough heat to boil water, which in turn makes steam. The steam is collected under high pressure and used to turn a turbine, which then spins an electromagnet. The spinning of the electromagnet produces a current of electrons. This current of moving electrons is what we refer to as *electricity*.

    i.   For fossil fuel-generated electricity, burning the fuel heats the water.

    ii. In nuclear power plants, the breaking apart of atomic nuclei produces the necessary heat.

Fossil Fuels

A. The majority of commercial energy is provided by coal, oil, and natural gas. They are called *fossil fuels* because they were made from the compressed carbon left behind by organisms that lived millions of years ago.

B. Coal

1. Coal is a sedimentary rock made from the remains of the trees and other plants of ancient forests. Those plants, through the process of photosynthesis, captured light energy from the sun and converted it into chemical energy in coal. Compared to other nonrenewable sources of energy, coal is relatively abundant and inexpensive.

2. Uses of coal. The primary use of coal in the United States and most other MEDCs is to generate electricity in large, coal-fired power plants. Coal can also be used in individual homes for heating. It can also be converted into coke, which is needed for steel manufacturing.

3. Formation of coal. Over time, heat and pressure increase and the moisture content of compressed plant material decreases, a process that leads to the eventual formation of coal. Classes of coal (listed in order of age and purity) include peat, lignite, bituminous coal, and anthracite.

   i.   Peat is the first step in coal formation and is not technically coal. Carbon-rich soils concentrate in bogs and fens and may be harvested to burn for fuel.

   ii. Brown and soft, lignite has a low carbon content of only about 30 percent. Because of this and its high water content, lignite produces less energy when burned (and more air-polluting emissions) than higher grades of coal.

        ➤ Almost half of the world's proven coal reserves are lignite, which is used primarily for electricity generation in power plants.

   iii. With a higher carbon content of 45 to 85 percent, bituminous coal is commonly used to generate electricity or it is converted into coke, which is necessary for steel making.

   iv. Anthracite is the purest form of coal, up to 97 percent carbon. This is the cleanest-burning coal due to low levels of impurities like sulfur and nitrogen.

        ➤ Anthracite is rarely found compared to other types of coal. As a result, it is the most expensive.

> ➤ Because of its lower emissions, anthracite is commonly used for home heating, and it is mixed with lower grades of coal in electricity-generating power plants.

4. Extraction. Coal is commonly mined in open-pit mines or strip mines (often resulting in mountaintop removal), although there are also underground coal-mining operations.

5. Pollution. To release its energy, coal is burned at high temperatures in order to heat water. The heated water evaporates into steam, which is used to turn turbines connected to electromagnets, which then produce an electric current. Burning coal is a major source of air pollution in MEDCs. Here is a list of some of the pollutants released into the atmosphere as a result of burning coal to produce electricity.

   i.   Carbon Dioxide (CO_2). Coal-fired power plants are the major source of anthropogenic carbon emissions. The CO_2 released by coal-fired power plants is a major contributor to global climate change, far more than car exhaust systems or deforestation.

   ii. Particulate Matter. Small particles of partially combusted ash are released from power plants. Particulate-matter pollution leads to respiratory problems, like asthma and chronic bronchitis, and contributes to smog formation.

   iii. Sulfur Oxides (SO_x) and Nitrogen Oxides (NO_x). SO_x and NO_x are also by-products of combustion. In the atmosphere, they combine to form sulfuric and nitric acids, which lead to acid deposition.

   iv. Other Toxins. Other toxins, including carbon monoxide (CO), arsenic, lead, cadmium, mercury, and trace amounts of uranium and thorium, are either released into the atmosphere or concentrated into waste ash. If waste ash is not properly treated and stored, these chemicals can leach into the environment, causing contamination of local freshwater (including groundwater).

6. Clean coal technology. Wet scrubbers may be installed in smokestacks to reduce emissions of sulfur. Electrostatic generators use static electricity to capture particulate matter.

Low NO$_x$ burners control the levels of oxygen in the coal fire to reduce nitrogen emissions. Some newer power plants use a combustion method called *limestone fluidized bed combustion*, which greatly reduces SO$_x$ and NO$_x$ emissions while also increasing the efficiency of the heat transfer. This generates more energy per unit of coal.

7. Coal reserves. More than 50 percent of the world's coal reserves are located in China, the United States, and Russia (including the former Soviet republics). It is estimated that, at current consumption rates, the world's coal reserves will be depleted within 150 years.

C. Oil/Petroleum

1. In its naturally occurring state, crude oil is a thick, sticky liquid that is formed from the remains of oceanic organisms that lived millions of years ago. Over time, these remains were heated and compressed to form oil. Once it is extracted, crude oil is refined to make a variety of different types of fuels and products.

2. Uses of oil. Although we think of oil mostly as a source of energy, it is also used to manufacture a variety of products we use every day.

   i.   Fuels. Some common fuels refined from crude oil include gasoline, diesel fuel, propane, and jet fuel. These fuels are separated in a distillation process called *fractioning*, in which they settle out based on density and boiling point into different chambers where they can then be collected and further treated to create useful fuels.

   ii. Products derived from oil. In addition to fuels, a wide variety of other products are made from the by-products of the refining process. These include nylon, plastics, asphalt, polystyrene, DDT, and even crayons!

3. Extraction. Oil is found in underground deposits in sites that were, at one time, along the ocean floor. To extract oil, deep wells are drilled and pumps are used to bring it to the surface. Oil wells may be accessed by building huge platforms at ocean sites, or the wells may be on land.

4. Oil Reserves.

    i.   Worldwide. The majority of the world's oil is located in the Middle East (Saudi Arabia, Iran, Iraq, Kuwait, the United Arab Emirates, and Qatar). There are also significant reserves in South America (Venezuela and Brazil), Africa (Libya and Nigeria), North America (Canada and the United States), and Russia.

    ii.   The United States. In the United States, the majority of proven oil reserves are located in Texas, Louisiana, California, the Gulf of Mexico, North Dakota, and Alaska.

5. World Oil Consumption. At the current rate of use, the world's oil reserves are expected to reach economic depletion within the next fifty years. World oil consumption doubled in the 1960s and 1970s due to increased use of automobiles. Another big increase is expected as India and China continue to urbanize and the use of cars in these countries increases. This will put even more pressure on oil reserves and continue to drive oil prices higher.

D. Natural Gas

1. Like coal and oil, natural gas is formed from the heated and compressed remains of ancient organisms.

    i.   Natural gas deposits are often found with deeper oil deposits, and when natural gas is found alone it is often located more than 2 miles below the Earth's surface.

    ii.   Natural gas is composed primarily of methane, although it also contains propane, ethane, and butane in smaller amounts.

2. Uses of natural gas. Because it is relatively free of impurities, natural gas is the cleanest-burning fossil fuel.

    i.   It is commonly used in home applications like heating, cooking, and powering appliances.

    ii.   About a quarter of the U.S. energy consumed comes from natural gas.

3. Extraction, storage, and transport. Natural gas is not perfect. There are serious environmental effects associated with the extraction, storage, and transportation of natural gas.

   i. Although burning natural gas releases fewer pollutants like SO_x, NO_x, and carbon dioxide, natural gas leaks are composed of almost pure methane. *Pure methane* is an aggressive greenhouse gas in its own right—about twenty times greater than carbon dioxide!

III. Nuclear Energy

A. *Nuclear energy* is the energy released by breaking apart the nucleus of an atom.

   1. The most common element used to generate nuclear energy is uranium, specifically an isotope called U-235.

   2. When a neutron particle is fired at high velocity at the nucleus of a U-235 atom, the collision causes the nucleus to break apart, releasing a great deal of energy.

   3. Nuclear energy is considered a nonrenewable resource because there is a limited amount of U-235 in nature.

B. Nuclear power plants are complex feats of engineering. Because the nuclear reaction used to generate electricity is so powerful and potentially dangerous, standards for design, construction, and maintenance are, by necessity, very rigid. This makes nuclear energy one of the most expensive ways of generating electricity.

C. Radioactivity

   1. Human health risks. Uranium is a radioactive element. Radioactivity is harmful to living organisms. As radioactive elements decay, they emit alpha, beta, and gamma particles. These particles can severely disrupt the DNA inside living cells as they pass through. As a result, humans and other organisms exposed to high levels of radiation suffer a variety

of devastating physical effects called *radiation sickness,* often resulting in a painful death. Even low levels of radiation can lead to an elevated risk of a variety of different cancers.

2. Nuclear waste. One of the major concerns people have with the use of nuclear power to generate electricity is how to store and dispose of nuclear waste safely.

   i. The uranium used to generate electricity in the plant has to be replaced every few months. When this occurs, the spent fuel rods are removed from the reactor and taken for storage. They are still highly radioactive even though they are no longer used in the reactor.

   ii. In the United States, nuclear waste is stored at the plant where it is generated. Depending on the level of radioactivity, waste may be buried in special concrete containers, stored in large pools of water (to ensure it remains cool), or placed in specially designed under-ground containment facilities.

3. Half-life. All radioactive elements have a unique *half-life,* the time it takes for half of the atoms in a sample to decay to a more stable form. For example, the carbon-14 isotope has a half-life of about 5,700 years. For a sample of 100 atoms of C-14, after 5,700 years, there will be 50 atoms of C-14 remaining in the sample. After another 5,700 years there will be 25 C-14 atoms left; after another 5,700 years, ap-proximately 12.5 atoms remain. This goes on infinitely, with half of the remaining atoms decaying at each passage of the half-life. We measure remaining radioactive atoms or radiation emitted by these atoms when we look at half-life calculations.

Test Tip

You are almost guaranteed to see a half-life calculation in some form on the AP exam. These calculations often appear in the multiple-choice section. Don't make the common mistake of thinking that, if 50 percent decays after one half-life, the remaining 50 percent will be decayed after the second pas-sage of half-life. It is an exponential decay, and half of what remains is lost at each step.

D. Nuclear Accidents

1. When a nuclear power plant is operating as it should, it releases very little pollution into the environment. Nuclear power is controversial because of what happens should the plant malfunction.

   i.   A major malfunction is often called a *nuclear meltdown*. The process of breaking uranium atoms apart releases huge amounts of energy. Uranium is a dangerous radio-active element.

   ii. To keep these dangers contained, nuclear power plants have extensive systems for cooling the reactive core and containing the nuclear reaction. There have been a few instances where these systems failed, and with devastating consequences.

1. Three Mile Island, Penn. (1979)

   ➤ The Three Mile Island (TMI) nuclear power plant experienced a series of failures in the mechanical operation, design, and human communication in the plant, which led to a partial meltdown of the reactor core.

   ➤ Although no one was injured or made ill by the incident, this was the first time Americans came to terms with the reality of how dangerous and devastating a nuclear meltdown could be, and how important it is to have a solid disaster management plan in place ahead of time.

   ➤ Many of the regulations and power plant designs in use today came about in response to the TMI event.

2. Chernobyl, Russia (1986)

   ➤ Due to a design flaw and operator error, the Chernobyl nuclear power plant experienced a total meltdown of its core and released a significant amount of radioactive gasses into the environment.

   ➤ To date, roughly 4,000 deaths have been attributed directly to the accident, a few from acute radiation poisoning soon after the accident, but the vast majority from thyroid cancer resulting from the ingestion of contaminated milk.

➤ More than 5 million people are considered to have been contaminated by radiation and are living with its effects, including cancer patients, children with severe birth defects, and people who have lost limbs from radiation exposure.

➤ To date, Chernobyl is the most devastating nuclear disaster that has occurred.

3. Fukushima Daiichi Nuclear Power Plant, Japan (2011)

➤ In the wake of a magnitude 9.0 earthquake and subsequent tsunami, three reactors at the Daiichi power plant overheated, and an explosion occurred.

➤ This event has been dubbed the worst nuclear meltdown since Chernobyl, and the full extent of damage, human health effects, and environmental impact will not be known for several years.

E. Pollution

1. Zero emissions. Unlike fossil fuels, nuclear power does not depend on combustion (burning) to generate energy. As a result, it produces no emissions of greenhouse gases in its operation.

2. Radioactive Waste. All power plants generate radioactive wastes, which can be harmful for thousands of years. Because of this, proper storage of this waste must be designed to withstand great passages of time and minimal human intervention to ensure its safety.

3. Thermal pollution. Water is essential to nuclear power plants because it is used to circulate around the reactor as a cooling agent. As a result, nuclear power plants are generally located near large bodies of water that may be drawn into the plant and then released back into the environment after circulating through the plant. If water is released directly to the environment without first being allowed to cool down, the high-temperature water may be harmful to organisms living in the body of water.

Math Practice

After 80 million years have passed, only $\frac{1}{128}$ *of the original number of radioactive atoms of an element in a sample remains. What is the half-life of the element?*

ANSWER:
Start with the proportion $\frac{1}{128}$. This is a passage of 7

half-lives of the element ($\frac{1}{2}$, $\frac{1}{4}$, $\frac{1}{8}$, $\frac{1}{16}$, $\frac{1}{32}$, $\frac{1}{64}$, $\frac{1}{128}$).
Next, divide 80 million years by 7 to get your answer: 11.4 million years.

REFERENCES

EIA Energy Kids-Coal. *http://www.eia.doe.gov/kids/energy.cfm?page=coal_home-basics*

EIA Energy Kids-Oil (petroleum). *http://www.eia.doe.gov/kids/energy.cfm?page=oil_home*

EIA Energy Kids-Natural Gas. *http://www.eia.doe.gov/kids/energy.cfm?page=natural_gas_home-basics*

How much coal is left: Energy Explained, Your Guide to Understanding Energy. *http://www.eia.doe.gov/energyexplained/index.cfm?page=coal_reserves*

The CIA World Fact Book. *https://www.cia.gov/library/publications/the-world-factbook/index.html*

NaturalGas.org. *http://www.naturalgas.org/overview/background.asp*

Institute for Energy and Environmental Research, Uranium Fact sheet. *http://www.ieer.org/fctsheet/uranium.html*

Renewable and Perpetual Energy Resources

 I. Solar Energy

A. Renewable resources can be renewed or regenerated (such as trees and forests) or they can be recycled infinitely (water). A perpetual resource is one that cannot be exhausted by human activities regardless how much of it we use. Solar energy is one example of a perpetual resource.

B. Solar energy is actually the ultimate source for all other types of renewable and many nonrenewable energy sources.

1. Heat from the sun drives the water cycle by causing evaporation and transpiration, thus making hydroelectric power possible.

2. Differences in air temperature and pressure caused by uneven heating of Earth's surface generate winds.

3. Light from the sun makes photosynthesis possible, which in turn makes it possible for biomass to accumulate.

4. Even fossil fuels were ultimately made possible from the sun's energy.

C. Active Solar Energy

1. *Photovoltaic cells* are used to collect and convert energy from the sun directly into electricity.

2. These collection cells are often mounted on the rooftops of houses and buildings; however, large-scale industrial solar arrays may cover acres of land (often in desert areas).

D. Passive Solar Energy

1. By using the sun's energy directly to accomplish a task, passive solar heating allows for a highly efficient source of energy.

    i. No mechanical, photovoltaic, or chemical conversions are used in passive solar energy.

    ii. Passive solar energy is most commonly used for heating systems.

2. Building design. A greenhouse is a good example of passive solar heating. The clear panes of glass allow the sun's rays to penetrate and warm the air inside, but the longer wavelengths of the infrared radiation (heat) cause it to be trapped inside, warming the greenhouse far beyond outside temperatures.

    i. By designing buildings with windows facing the east, west, and south, the sun's capacity for heating a building can be maximized.

    ii. Using dark colors in areas where heat is desired increases the effectiveness of heat absorption. For example, using dark-colored shingles can reduce heating bills in northern climates.

    iii. Conversely, planting shade trees, using reflective window coverings, and using light colors for roofing materials can greatly reduce cooling expenses in warmer climates by reducing a building's heat absorption potential.

3. Passive solar heating. Passive solar heating can be used to generate electricity by heating water to make steam that turns turbines. In these systems, a series of mirrors are used to concentrate sunlight in order to heat the water efficiently.

E. Benefits of Using Solar Energy

1. Zero emissions and pollutants. Because nothing is burned (at least, not on Earth!) to produce electricity, all of the combustion-related pollutants are avoided by using solar power.

2. Habitat preservation. No mining or transport of fuel is required. As a result, harmful practices such as mountaintop removal and pit mining are avoided. Because many solar applications can be installed directly on rooftops in already developed areas, animal habitats are not disturbed.

3. Efficiency. Passive solar heating is extremely efficient because of the simplicity and minimization of energy conversions. It also requires a low initial monetary investment, especially when passive solar design elements are incorporated into the plans of a new construction project.

F. Drawbacks of Using Solar Energy

1. Photovoltaic cells are complex pieces of equipment that must be manufactured. This means that materials must be mined or extracted; factories must be built (which use building materials and energy); and the process of manufacturing produces its own host of toxic chemicals, which contribute to atmospheric and water pollution.

2. Photovoltaic cells have a finite lifespan and must be replaced. This contributes to the solid waste disposal problem. It also means that more materials and energy must be expended to produce the replacement equipment.

3. Batteries are often used to store excess energy for use during times when the sun is not out. Manufacture and disposal of these batteries is another environmental concern, especially because modern long-life batteries use heavy metals like cadmium, lead, and mercury. All of these heavy metals are considered powerful toxins when they are released into the environment by improper disposal.

II. Hydroelectric Power

A. Benefits of Building Dams

1. People have been using dams to redirect water and do work for centuries. Dams allow us to direct water toward our crops for irrigation, control the annual flooding cycle of large rivers, and produce electricity.

2. Energy generation

   i.   Large rivers, like the Colorado River in the western United States, give us the opportunity to build hydroelectric dams. These dams control how much water flows down the river and funnel the flowing water through the dam to produce electricity.

   ii.   Behind the dam, a reservoir, or large human-made lake, is created by the water that is held back. This system allows a constant stream of water to flow through the dam, producing a reliable supply of electricity with no emission or chemical by-products.

3. Predictable water supply and flood control

   i.   Rivers with no dams often have a seasonal fluctuation of water. In spring, torrents of water are released as snow and ice melts in the mountain source areas. In addition, spring rainfalls often lead to flooding. In winter, the weather is drier, and most precipitation falls as snow and ice in the mountains, so the flow of the river is greatly decreased.

   ii.   Capturing a portion of river water in a reservoir behind the dam creates a year-round source of water, which can be used to irrigate crops and supply water to homes and businesses.

4. Recreation

   i.   The reservoirs that accumulate behind large dams become huge and often beautiful lakes. Many forms of recreation such as fishing, boating, and swimming are popular in these areas.

B. Negative Effects of Dams

1. Water Diversion. Aside from producing electricity, dams are also useful for changing the route of the flow of water, known as *water diversion*.

   i.   Aral Sea, Former Soviet Union

       ➤   An ancient inland saltwater sea, the Aral Sea was one of the world's largest lakes and a source of fishing.

➤ Due to water diversions resulting from Soviet-era dams built on the two large rivers feeding into the Aral Sea, it has lost 60 percent of its total surface area.

➤ The dams were built to provide irrigation water to large-scale monoculture cotton farms.

➤ In addition, increased salinity and concentrations of toxins from agricultural runoff have reached the point where all of the species of organisms that lived in the sea have been driven to extinction. The sea is now lifeless.

ii. James Bay, Quebec

➤ The James Bay Project (JBP) consists of a series of hydroelectric dams that were built beginning in the 1970s to provide electricity to Quebec, Canada.

➤ Controversy has surrounded the JBP because the flooded areas and water diversions have had a significant negative impact on the population of Cree Indians who live in the area.

➤ Altered river flow and changes in ice formation have decreased hunting opportunities, and lands formerly used by the Cree for hunting and living have been flooded.

2. Reduced Fertility of Floodplains

i. The annual cycle of water flow. While inconvenient to humans, the natural cycle of flooding and drought common to large rivers is an important part of maintaining soil fertility in land adjacent to the river.

➤ It is no coincidence that the most successful of our ancient agricultural civilizations were built along large rivers like the Nile, Tigris, and Euphrates. As river water floods over the banks, the fast-moving water carries nutrient-rich sediments that are deposited on the flooded land as the water slows and settles. When the water recedes, the sediments remain, enriching the soils and making them more productive for growing plants.

ii. Dams alter the cycle. Dams that prevent flooding tend to accumulate sediments upstream of the dam and deprive the downstream areas of these nutrients.

iii. Disruption of aquatic organisms. In addition to changing the amount of water that flows in a river, dams also create a physical barrier to organisms attempting to navigate along the river.

➤ Salmon are a commercially important migratory fish that are unable to complete their life cycle and breed when they are blocked by a dam. Some dams incorporate "fish ladders," which are a type of stream flowing over the top or around the side of the dam to facilitate migration of fish and other aquatic life.

C. Hydroelectric Power Case Studies

1. Colorado River. The Colorado River and its tributaries flow through the southwestern United States. More than 20 dams have been built on the Colorado River, including the Hoover Dam. While hydroelectric power is generated at many of these dams, the primary purpose for most of the Colorado River dams is to divert water for irrigation in the very arid states of the southwestern United States.

2. Three Gorges Dam, Yangtze River, China. The Three Gorges Dam is either one of humankind's greatest engineering feats or most devastating ecological disasters. It all depends on your perspective.

i. Three Gorges Dam is the world's largest hydroelectric dam, which makes it a great engineering feat.

➤ The power it provides to China greatly reduces the need to burn fossil fuels. Not only is it a zero emissions electricity generator, it also reduces the fossil fuel consumption needed to mine and transport coal.

> ➤ In addition to providing electricity, the dam brings other benefits, like flood control, downstream. (Yangtze River floods mean thousands of human deaths and millions of dollars in damage.)

> ➤ The dam has also increased shipping capabilities because of the deeper and wider channel created by the flooded regions behind the dam.

   ii. But some see the dam as an ecological disaster. The area flooded by the construction of Three Gorges Dam created a reservoir 360 miles long.

> ➤ The reservoir displaced more than 1 million people as it inundated cities, towns, factory sites, farmland, and areas of cultural and archaeological significance.

> ➤ The excess water upstream of the dam has also led to increased landslides and erosion.

> ➤ Because much of the flooded area was developed, toxins from factory residues, pesticides, and other chemicals continue to leach into the water.

D. Ocean/Tidal Energy Generation

  1. More than 70 percent of Earth's surface is covered by ocean, so it only makes sense that we would explore ways to use ocean water to generate power. There are two major ways to do this: collecting thermal energy produced by solar warming ocean water and harnessing the power of moving ocean water.

  2. Thermal energy to electricity.

   i. Water has a high heat capacity. As the sun shines down on the oceans, the water absorbs and stores a proportion of that radiation as heat. This explains why surface ocean water is much warmer than deep ocean water.

   ii. It is possible to use the heat in ocean water to evaporate low boiling-point liquids, such as ammonia, to drive turbines and produce electricity.

3. Tidal energy. The gravitational attraction between the Earth and the moon produces the tidal movement of water.

   i.   Differences in water temperature caused by the uneven heating of Earth combined with surface winds also contribute to water currents, upwellings, and other movement of water in the oceans.

   ii. Tidal energy collectors work in a manner similar to hydroelectric dams to harness the power of moving water in order to generate electricity.

III. Wind Power

A. People have been harnessing the power of the wind to produce energy for centuries. The original wooden windmills that we associate with early rural farm life are really not very different from the high-tech wind turbines built today. The basic concept remains unchanged—long blades catch the wind and rotate, converting wind energy into mechanical energy that can be used for pumping water, grinding grains, or generating electricity.

B. Benefits of Wind Power

   1. No emissions. Again, because nothing is being burned, no atmospheric emissions are released in the energy-generating process.

   2. Simplicity and safety. Wind turbines are relatively simple to construct, require little maintenance, and are a reliable source of energy when properly located. At worst, a wind turbine may leak some lubricating fluid or produce an annoying sound as it spins. Compared to a nuclear meltdown, windmills look pretty safe!

   3. Small land footprint. A wind turbine is a tall structure that takes up a small piece of land. This allows multiple uses of wind-farm land. For instance, it is common to graze livestock on land that is also being used as a wind farm. Wind turbines are often located offshore; the strong sea breezes

generated over shallow water are perfect for powering a turbine.

C. Drawbacks of Wind Power

1. Manufacturing the turbines. Most modern wind turbines are made of metals and plastics, so there are negative environmental effects associated with extracting or manufacturing the materials needed for their construction.

2. Spinning blades. Because energy is generated by large spinning blades, aerial organisms like birds, bats, and migrating butterflies can be fatally injured by contact with the blades. This is particularly a problem for bats because they navigate primarily by sonar, which is unable to detect the position of the windmill's spinning blades.

IV. Biomass

A. Anyone who has ever roasted marshmallows over a campfire has experienced biomass energy. Burning wood or charcoal is a simple example of generating power through biomass.

B. Biomass in Less Economically Developed Countries (LEDCs).

1. LEDCs have less infrastructure for large-scale energy generation. Many people rely on wood- and charcoal-burning stoves for heating and cooking. Unfortunately, this great demand for wood is a major cause of deforestation, and the incomplete combustion of burning wood at relatively low temperatures releases high levels of soot, greenhouse gases, and other pollutants into the atmosphere.

C. Biomass in More Economically Developed Countries (MEDCs)

1. In some MEDCs, such as the United States, attempts are being made to decrease the need for fossil fuels and look for more sustainable sources of energy. Large-scale biomass power plants are being built. At the same time, technological advances in clean-burning wood pellet stoves are encouraging more people to use them for home heating.

D. Types of Biomass Power

1. Biodiesel

    i. As an alternative to using petroleum fuel for automobiles, some people are turning to *biodiesel*, which is a fuel made from plant oils and other organic materials.

    ii. In the United States, corn is commonly converted into ethanol, which can be used alone or mixed with petroleum oil and used to fuel cars without requiring any kind of engine modifications.

        ➤ Other plants like soybeans, safflower, and sunflower can also be used to produce oils that can be converted to biodiesel fuel.

        ➤ There is a great deal of current research devoted to using algae to produce oils that can be burned for energy.

        ➤ To power a car purely on biodiesel fuel, some modifications must be made to the car.

    iii. Another source of biodiesel fuel is used cooking oil. Some restaurants gladly give used cooking oil to owners of biodiesel cars, saving themselves the cost of disposing of the oil and providing car owners with free fuel.

    iv. Like other fuels, biodiesel has its problems. First, because it is burned to produce energy, toxic chemical emissions like carbon monoxide and greenhouse gases like carbon dioxide are still produced.

2. Waste to Energy (WTE)

    i. WTE can be as simple as burning trash. Although WTE is not a very efficient or inexpensive way to produce energy, it does solve two problems at the same time: Energy is produced while solid waste is incinerated instead of accumulating in landfills.

    ii. WTE makes the most sense in areas where sources of fuel and landfill space are limited.

3. Biogas

   i. Methane is commonly produced from the decay of organic materials.

   ii. In landfills, methane is produced within a pile of decaying garbage. Because methane is an aggressive greenhouse gas, it makes good sense to vent methane from the landfill, collect it, and use it for fuel. Remember, natural gas is primarily composed of methane.

   iii. Large-scale farming operations produce a great deal of animal wastes. Manure ponds or mechanical digesters may be used to break down these wastes. The resulting methane is collected and used as a source of energy. This process is called *anaerobic methane digestion*.

V. Geothermal Energy

A. *Geothermal energy* is energy from the Earth.

B. Geology Review

1. If you recall the theory of plate tectonics, the interior of the Earth is divided into layers of rocks and minerals that vary by temperature and density, basically becoming warmer as one approaches the innermost core. Just below the surface crust, large deposits of superheated rock called *magma* form primarily as a result of the radioactive decay of atoms in the Earth's core. Earth's crust is made up of large tectonic plates that slowly move across the aesthenosphere. They are driven by convection currents within Earth's mantle. At the boundaries where these plates interact, the crust tends to be thinner and allow for this heat to rise closer to the surface of the crust, often resulting in volcanoes.

C. How Geothermal Energy Is Converted to Useful Energy

1. The simplest way to capture geothermal energy is to dig a pipeline down into the crust to a hot spot and send water down to be superheated, then returned to the surface as

steam to power a turbine. As the steam cools and condenses back to water, it is sent back underground again to continue the cycle.

VI. Automotive Alternatives to Fossil Fuels

A. Electric and Hybrid Cars

1. All gasoline-powered cars have a battery. The internal combustion engine burns fuel to power the drive train of the car and to keep the battery charged to run all the electrical components of the car. In an electric car, the battery does all the work. The battery is charged between uses by plugging it into an electrical power source. Hybrid cars work on the same principle, but they also have a smaller internal combustion engine to use as a backup to extend the range of the battery. At present, battery technology cannot produce a battery with a long enough life to allow for long car trips without frequent recharging.

2. Emissions. Electric cars produce no emissions as they operate, and emissions are greatly reduced in hybrids, but these cars still rely on electricity, which is often produced by coal-fired power plants to recharge the battery. Basically, the emissions benefit is that pollution is transferred from many, smaller non–point sources (individual vehicles) to a single, larger point source (the power plant).

3. Batteries. As with wind power, the batteries used to store electricity have a finite life span. In addition to being extremely expensive to replace, they are often harmful to the environment if disposed of improperly.

4. Inconvenience. With current automotive technology, battery range is the greatest concern. Gasoline is readily available, and filling a gas tank is a quick and familiar process. Batteries currently used in electric cars have a limited range, and recharging takes about 30 minutes to reach 80 percent capacity. Recharging stations are not common, although they are increasingly available in Japan, and the United States has

plans to install charging stations throughout the nation in the near future.

B. Hydrogen as Fuel

1. Pure hydrogen gas is not found in nature. To produce hydrogen for fuel, hydrogen atoms must be removed from hydrogen-containing molecules. One process, *electrolysis*, uses an electric current to separate the hydrogen and oxygen atoms in water.

     i.   One of the most significant uses of hydrogen power is in NASA's solid rocket boosters of the space shuttles. Hydrogen fuel cells within the shuttles supply electricity inside the shuttle and provide drinking water for the crew as a by-product.

     ii. Brazil and Australia have invested in hydrogen fuel cell–powered buses for public transportation.

VII. Local Generation Versus Centralized (Grid) Generation

A. Local Generation

1. Placing a solar collector on the roof of a house or locating a wind turbine in the backyard allows a homeowner to produce the electricity needed in that house and reduces dependency on a centralized power grid. This gives individuals the option to make choices about their environmental impact on a person-by-person basis. In many cases, it saves money, too, because no power bills would be owed to an electric company. It also makes it possible for people in LEDCs, where no power grid is available, to have access to electricity. However, the cost of installing and maintaining the necessary equipment may be high, and there is no guarantee of continuous and reliable power.

B. Large-Scale Power Generation ("The Grid")

1. In most MEDCs, there is a well-established municipal power grid that supplies a continuous and affordable source of

electricity to homes and businesses. The electricity is usually generated at a large power plant, hydroelectric dam, wind farm, or solar field. This is convenient because an individual does not need to think about where energy is coming from; it is reliable as long as the monthly utility bills are paid. The drawback is that some methods of providing this electricity are environmentally damaging, and for some it is a moral compromise to use energy generated in an unsustainable fashion. People who get their energy from centralized power corporations are also subject to pay the prices set by the power company and have few opportunities to shop around for the best price or product.

When thinking about the relative merits of each type of energy, a few rules hold true across them all:

1. *There is no perfect source of energy. The best way to reduce our environmental impact is to use less energy, period. Alternatives such as solar power and wind power require harmful mining and extraction processes to obtain the raw materials needed to make the necessary equipment or machinery.*

2. *Regardless of where it originates, any fuel that must be burned will produce atmospheric pollution.*

3. *Renewable energy does not automatically equal environmentally friendly energy. For instance, some types of biomass fuels produce far more pollution and negative environmental effects than some nonrenewable sources.*

REFERENCES

EIA Energy Kids—Biomass. *http://www.eia.doe.gov/kids/energy. cfm?page=biomass_home*

Ocean Energy—Renewable Energy World. *http://www.eia.doe. gov/kids/energy.cfm?page=biomass_home*

Environmental Toxicology and Pollution

I. Toxicity

A. Measuring Toxicity

1. One way to measure the toxicity of a substance is to test the effects of exposure on nonhuman organisms. Whether or not you find animal testing ethical, the reality is that every medicine approved by the U.S. Food and Drug Administration (FDA) must be first tested on animals to determine its possible harmful side effects.

2. LD_{50} is the amount of a substance that will be a lethal dose to 50 percent of a test population. Usually measured in mg of substance per kg of body weight, the lower a substance's LD_{50} value, the greater its toxicity (meaning a smaller dose will be fatal). Because it is measured in mg per kg of body weight, we can extrapolate values obtained on small mammals like rodents to estimate a range of lethal doses for humans, although this is not always an accurate estimation because of physiological differences between different species.

3. LC_{50} is the lethal concentration that will kill 50 percent of a test population. This measurement is used with aquatic organisms and refers to the concentration of a toxin dissolved in water that will kill organisms living in that water.

4. Measuring Amount of a Substance

   i. When measuring levels of a toxin in the environment, even very small amounts may often have harmful environmental effects.

      ii. Parts per million (ppm) and parts per billion (ppb) are units commonly used to measure the amount of a substance in the environment. For example, in a sample of water, a nitrate level of 10 ppm is considered unsafe for drinking. This means that there are 10 parts of nitrate per million parts of water. In liquids, 1 ppm is equivalent to 1 mg/L.

B. Acute Versus Chronic Toxicity

1. *Acute toxicity* is a measure of the immediate, short-term effects of exposure to a toxin. If you ingest meat tainted with salmonella bacteria, you may become very ill within a few days and possibly die. If you survive, however, you will start to feel better as your immune system fights off the bacteria and its toxic effects decrease until eventually you feel healthy again.

2. *Chronic toxicity* refers to a physiological effect that will not be immediately apparent but will show up several months or even years after the exposure. Chronic toxicity is also seen in situations where people are exposed to low, seemingly harmless doses of a toxin over a long period of time.

    ➤ For example, it was recently reported that many of the plastics used for water bottles emit a chemical called BPA. While we do not expect to see anyone dropping dead after drinking out of a plastic water bottle, the data suggests that people, especially babies and small children, who use these water bottles frequently over long periods of time have an increased risk of negative health effects like cancer, neurological dysfunction, and even obesity.

C. Synergy

1. In some cases, two or more seemingly harmless substances have a toxic effect in combination. Another way to describe this is "the whole is greater than the sum of its parts."

2. Synergistic toxicity is of particular concern in aquatic ecosystems where a variety of pesticides, heavy metals, and other toxins accumulate from industrial, roadway, and agricultural

runoff. Although government agencies have set toxicity limits for many individual chemicals, the synergistic effects of many of these in combination are unknown.

D. Threshold of Toxicity

1. The *threshold of toxicity* is the limit below which no harmful effect is seen. Once the toxic threshold is reached or exceeded, harmful effects begin to appear.

   ➤ For example, a blood level of 0.08 percent concentration is widely held in the United States as the toxic threshold for human alcohol consumption. Above that level, it is illegal to drive a car because of physical and mental impairment caused by the acute effects of the alcohol on your body.

   ➤ People who work in nuclear power plants must wear a badge that measures their daily exposure to background radiation in the plant. This is to ensure that no employee reaches the threshold of toxicity for radiation exposure. If an employee's exposure approaches that level on any given day, he or she is sent home.

E. Bioaccumulation and Biomagnification

1. *Bioaccumulation* is the process of a toxin or other chemical building up in an organism over time and with multiple exposures.

   ➤ For example, bracken ferns are a fast-growing, hardy plant that, when planted in soils contaminated with toxins such as arsenic, can take in the contaminant through their roots and store it in their leaves and other tissues. When harvested and tested, the levels of arsenic in the tissues of the ferns far exceed the levels in the surrounding soils.

   ➤ For many organisms, bioaccumulation leads to chronic toxicity effects like weakened immune systems and neurological dysfunctions.

2. *Biomagnification* is the process where an accumulated toxin is passed up the food chain as one organism eats another.

➤ For example, DDT is a pesticide that is very effective at killing mosquitoes. In low doses, DDT is not particularly harmful to larger organisms, so it was widely used throughout the tropics, the United States, and Europe. In the 1960s, scientists began to notice that large predatory birds were declining in record numbers because there was something interfering with the calcium in their eggs, leaving them soft and prone to breaking. When affected birds were tested, alarmingly high levels of DDT were present in their tissues.

➤ Here's how biomagnification in the case of DDT works:

— DDT was sprayed in the water to kill mosquito larvae. Small zooplankton and phytoplankton species accumulated small doses of DDT in their cells (one dose each, for example).

— Small fish eat plankton. Each time a fish eats a plankton, it acquires the plankton's one dose of DDT in its own body, in addition to any DDT it would take in merely by being exposed to the sprayed water. Let's say that each small fish has 10 additional doses of DDT from eating plankton plus 1 dose from living in the sprayed water, equaling 11 doses of DDT per fish.

— Large fish eat small fish, and lots of them. If a large fish eats 20 small fish, it also gets the DDT accumulated in the tissues of each small fish it eats, so that is 220 doses in addition to the 1 dose it receives from living in sprayed water, equaling 221 doses of DDT per large fish.

— Along comes an eagle, who catches and eats the big fish. For every fish the eagle eats, she gets a dose of 221 units of DDT. If she eats only 5 fish, that's a total of 1,105 units of DDT. Keep in mind that birds have a relatively low body weight for their size; it is no wonder they were being harmed by the small amounts of DDT in the environment.

➤ Now you see how even a low concentration of a toxin can wreak havoc on top predators in a food web due to biomagnification.

II. Pollution

Pollution may be defined as the addition of anything into a natural environment that may cause harm. Pollution is related to toxicity because the more toxic a substance, the greater the potential damage to the environment. Aside from toxicity, a few other factors help to determine how much damage a substance might do.

A. Persistence

Persistence is defined as how long a substance remains in an environment.

1. Measuring the half-life of a radioactive isotope, such as uranium 238, is an example of measuring persistence.

2. Another way of measuring persistence is to look at how long it takes for a substance to disintegrate. For example, if you toss a banana peel on the ground and come back several weeks later to look for it, the banana peel will likely be completely gone. Organic materials that can be readily decomposed by bacteria have a very low persistence in the environment.

3. If you drop a plastic bottle instead of a banana peel, that bottle will be relatively unchanged for a very long time. You could conceivably come back years later and find it looking very much the same—dirty perhaps, but a plastic bottle all the same. Synthetic materials like plastics have a very long persistence in the environment.

III. Point Versus Non–Point Sources of Pollution

A. Point sources of pollution are easy to identify. Literally, you can point to the smokestack of a factory and say, "That's where the pollution is originating."

1. Effluent pipes carrying industrial waste from factories into waterways used to be common point sources.

2. More recently, the Deep Water Horizon oil leak in the Gulf of Mexico in 2010 was a dramatic point source of oil pollution.

B. Non–point sources of pollution are more difficult to identify.

1. Runoff is a major non–point source pollutant. Agricultural runoff is water that accumulates nutrients like nitrates and phosphates from fertilizers or animal wastes, as well as toxic chemicals found in herbicides and pesticides. All these chemicals dissolve in the water and are carried to waterways.

2. Acid deposition is another non–point source pollutant. Nitric and sulfuric acids are formed in the atmosphere from NO_x and SO_x (respectively) emitted from power plants and industrial operations. The acid rain or other types of deposition that return to the ground and water may be miles away from the original emissions, and they may contain a mixture of acids formed from different airborne pollutants.

3. One characteristic common to all non–point source pollutants is that they rarely contain toxins from a single event or location. They usually contain a mix of contaminants from several sites that mix together in the environment.

It is important to see how many of the issues we study in Advanced Placement Environmental Science (APES) are inter-related. For example, we see that our choices of which energy sources to use will have a huge impact on the levels of pollutants. Sometimes the toxicity of potential pollution drives the choices we make. The Three Mile Island nuclear disaster totally changed the public outlook concerning the desirability of nuclear energy.

IV. Pollutants That Cause Problems Across Air, Water, and Land

A. Heavy Metals

1. Heavy metals like lead, mercury, cadmium, and cobalt are potent neurotoxins, and low-level exposures may be harmful. As technology continues to advance, we are encountering more and more heavy metals in our computers, cell phones, and batteries.

2. Improper disposal of items containing heavy metals can lead to leaching into the soil and groundwater.

3. Heavy metal pollution is even an atmospheric problem. Among the emissions created by burning fossil fuels are mercury, thorium, and other heavy metals.

B. Radioactive Wastes

1. Radioactive isotopes emit potentially harmful alpha, beta, and gamma waves that can enter cells and disrupt DNA, making exposure very dangerous. These waves move through the air, so direct contact is not necessary.

2. Soil contaminated with ash from a radioactive exposure can cause food plants grown in it to bioaccumulate harmful radiation, which may then be ingested and be even more dangerous.

3. Water exposed to a radioactive element will also become toxic and be potentially harmful to any organisms that come into contact with it.

C. Acids

1. Oxides of nitrogen and sulfur (NO_x and SO_x, respectively) are common air pollutants produced by industry, power plants, and automobiles. When these gases accumulate in the atmosphere, they react with water vapor to form nitric and sulfuric acids.

2. These air pollutants are the major culprits in acid rain and other types of acid deposition. When these acids enter soils or aquatic ecosystems, many of the organisms in those habitats are unable to tolerate the acidity and die off.

V. Reduction and Remediation

A. If you have ever tried to clean up a huge mess, you know that sometimes, no matter how hard you try, you can never restore things to their original state of cleanliness. Environmental remediation works the same way. Once pollutants have been released into the environment, it is very difficult to remove them. In some cases like oil spills, the process of trying to clean up the spill actually damages the environment almost as much as the oil itself. The most sensible approach is to do whatever is possible to prevent pollution from happening.

REFERENCES

Lethal Dose (LD50) Values/Ag 101/Agriculture/US EPA. *http://www.epa.gov/agriculture/ag101/pestlethal.html*

West Virginia University Extension Service: Agricultural Practices and Nitrate Pollution of Water. *http://www.caf.wvu.edu/~forage/nitratepollution/nitrate.htm*

How Human Activities Affect the Atmosphere

I. **Primary Sources of Air Pollution**

A. Air pollution is caused by anything added into the atmosphere that can have a negative effect on the health of ecosystems and the organisms that make them up (humans included).

B. Point Sources. Many air pollutants are added to the atmosphere by easily identifiable sources. A *point source* is a single, stationary, localized source of emissions. Examples include the smokestacks of factories or power plants, methane emissions from a landfill, or smoke and ash from a forest fire.

C. Non–Point Sources. Also known as dispersed or mobile sources of pollution, *non–point sources* are usually difficult to identify directly as the source for a particular pollution problem. For example, acid rain is a result of non–point air pollution. Even though some of the chemicals that combine to make acid rain may have originated from identifiable smokestacks, most pollutants, once released into the atmosphere, travel with wind currents where they mix and chemically recombine. This means that the ultimate source of pollution over a particular area may be hundreds of miles away and impossible to identify directly.

The major theme about air pollution is that combustion (burning) of organic material in any of its forms releases a host of chemicals into the atmosphere, many of which are toxic to living organisms. This includes fossil fuels, wood, and other biomass, even cigarettes! The majority of these pollutants aggravate respiratory tissues, resulting in increased rates of asthma, lung cancer, and other respiratory diseases.

C. Major Outdoor Primary Air Pollutants

Pollutant	Sources	Human/Environmental Effects	Reduction/Remediation
NO_x: nitrogen oxides, including NO, NO_2, and N_2O	A common product of combustion; found in auto emissions, power plant smokestacks, etc.	Causes respiratory irritation and asthma; combines in the atmosphere to produce tropospheric ozone (a major component of smog) and nitric acid, which may fall as acid deposition	Reduce dependence on combustion to produce usable energy; strict emissions standards for vehicle exhaust; holding coal-fired power plants to higher emissions standards, using low NO_x burners
SO_2: sulfur dioxide	Mostly from burning fossil fuels in coal-burning power plants and factories; volcanic eruptions	Respiratory irritant and aggravates asthma, can lead to bronchitis and emphysema; can combine in the atmosphere to form sulfuric acid leading to acid deposition	Reduce dependence on combustion to produce usable energy; burn higher grades of coal (such as anthracite), which contains less sulfur; control SO_2 in smokestacks by using wet scrubbers that utilize limestone or sea water; use fluidized combustion technology
Particulate matter (PM) of less than 10 μm in diameter	Burning any type of organic materials from fossil fuels to forest fires; dust and dirt from unpaved roads and construction sites	Particles can accumulate in and damage lung tissue; the smallest of these particles can enter a person's bloodstream; aggravates respiratory tissues, which can lead to asthma, bronchitis, emphysema; increases the likelihood of developing heart disease, lung cancer; forms industrial smog	Reduce dependence on combustion to produce usable energy; use electrostatic generators in power plants to capture particles before they leave the smokestack
CH_4: methane	A natural product of anaerobic decomposition (including gas from living organisms like cattle and termites); landfills; natural gas leaks	Twenty times more potent greenhouse gas than carbon dioxide; reacts with NO_x to make ozone in the troposphere	The most common way to reduce methane pollution is to trap methane and use it as a source of energy production (natural gas)

II. Secondary Air Pollutants

A. Secondary air pollutants form as a result of chemical reactions among primary pollutants in the atmosphere. As a result, the only way to prevent secondary pollutants is to eliminate the primary pollutants that cause them to form.

B. Acid Deposition. More commonly known as acid rain, acid deposition includes both wet deposition (rain) and dry deposition (aerosols and particles) of acids from the atmosphere.

1. Causes of Acid Deposition. Nitrogen oxides (NO_x) and sulfur dioxide (SO_2) are released into the atmosphere primarily by burning fossil fuels. They react with water and other chemicals in the presence of sunlight to produce nitric and sulfuric acids.

2. Environmental Effects of Acid Deposition

   i.   Acidic rainwater. The pH of rainwater is naturally acidic, about 5.6 pH. This is because rainwater naturally reacts with carbon dioxide in the atmosphere to form carbonic acid as it falls. In areas with high levels of industrial pollution, rainwater pH is lowered even further into the acid range as it reacts with sulfur dioxide (SO_2) to form sulfuric acid and nitrogen oxides (NO_x) to form nitric acid.

   ii. Disrupted ecosystems in lakes, streams, and rivers. Organisms in aquatic ecosystems each have a particular range of tolerance for pH. While some species may thrive in acidic environments, most species prefer water that is near to neutral pH.

   iii. Acidification of soil. Acidification leads to the leaching of essential nutrients from the soil. It may also lead to chemical reactions releasing excess aluminum into the soil. Aluminum is a metal that in elevated concentrations is toxic to most forms of life.

   iv. Corrosion of buildings, bridges, and public art. Acids are corrosive. Even small amounts of acid deposition may, over time, lead to corrosion of concrete, marble, and the metals used in buildings, bridges, and public art.

C. Tropospheric Ozone (O_3)

1. Although ozone forms naturally in the atmosphere, human activities can cause it to become more concentrated in the lower atmosphere. Ozone forms as a product of the reaction of nitrogen oxides with volatile organic compounds (VOCs), such as hydrocarbons in the atmosphere.

2. Ozone is a very corrosive chemical and can destroy plastics, fabrics, nylon, and living tissues. The irritation of plant tissues reduces their ability to perform photosynthesis.

3. Ozone is an ingredient in photochemical smog.

D. Photochemical Smog

1. *Photochemical smog* forms from a combination of air pollutants reacting with one another in the presence of sunlight.

2. Formation of photochemical smog. Reactions of NO_x and VOCs in the atmosphere lead to the formation of small particles of matter, which produce a haze in the air and obscure range of sight.

3. *Thermal inversions*. Photochemical smog may be intensified by a thermal inversion, where a column of descending cool air traps the smog close to the ground.

    i. Mexico City, Los Angeles, and Beijing are three cities known to have serious problems with this type of air pollution.

III. **Indoor Pollutants and Sick Buildings**

A. Sick Building Syndrome

1. As early as the 1980s, people discovered that certain buildings seemed to be making them sick. *Sick building syndrome* is a situation in which a large proportion of otherwise healthy people living or working in a building together suffer from respiratory complaints, allergic reactions, or other symptoms that seem to be caused by exposure to the building. Causes of sick building syndrome include lack of adequate ventilation and chemical and biological contaminants.

    i. Lack of adequate ventilation. There is a conflict between ventilation and efficiency of air conditioning systems. In some cases, the air in a closed building is not exchanged often enough to remove all of the contaminants that accumulate from people and activities inside the building.

    ii. Chemical and biological contaminants. Examples of chemical and biological contaminants are listed in the following table.

Indoor Air Pollutant	Sources	Human/ Environmental Effects	Reduction/Remediation
Mold, fungus, bacteria	These biological contaminants may come from the outside and grow in ventilation ducts, insulation, carpets, ceiling tiles, and so on, especially where heat and moisture are present.	This is a major cause of sick building syndrome; allergic reactions include headaches, coughing, sneezing, dizziness and respiratory irritation.	Proper maintenance of air filters; increased ventilation; in many cases, the only remedy is to remove and replace all materials that may be contaminated.
Asbestos	In buildings built before 1975, asbestos was commonly used in floor and ceiling tiles, insulation, and other applications.	Tiny, microscopic fibers that are breathed can cause lung cancer 20 or more years after exposure.	In 1975, the use of asbestos was banned; in older buildings, asbestos may need to be removed by professionals, or left in place and covered over.
Cigarette smoke	Cigarette smoke is actually more harmful to the nonsmokers breathing secondary smoke than it is to the smokers.	Increased risk of lung cancer; respiratory irritant that increases the risk of asthma, bronchitis, and emphysema.	Smoking inside public buildings, restaurants, and workplaces has been banned in many cities and states in the United States and throughout the world.
Radon	A natural decay product of U-238 in bedrock, it is found in almost all water, soil, and rock. It enters homes through cracks and opening in floors and walls.	Lung cancer (second only to smoking in causing lung cancer deaths).	Ventilation is the most reliable way to reduce radon in the home; special home ventilation systems draw radon up from the soil and send it out above the house.
Carbon monoxide (CO)	Improperly ventilated fireplaces; leaks from gas appliances like water heaters, stoves, and clothes dryers; the tailpipe of a car in a closed garage.	CO is an odorless, invisible gas that can kill you before you are even aware it is there. CO binds to hemoglobin in the blood, and keeps it from picking up and carrying oxygen to the body's cells; it essentially leads to suffocation.	Every home that has gas appliances and/or wood-burning fireplaces needs to have carbon monoxide detectors; the detectors need to be checked regularly. Gas appliances need to be professionally installed and vented.
Volatile organic compounds (VOCs)	VOCs are found in many products: new carpet, paint, foam padding in furniture, and household cleaners, to name a few.	Cause a wide array of harmful health effects, including eye and respiratory irritation, headaches and dizziness, increased risk of cancer.	Many products are now made without VOCs. Properly ventilate your house—open windows and doors to increase airflow. Properly store and dispose of chemicals, including paints and solvents.

IV. **How Cities Are Different**

A. An urban heat island (UHI) is a metropolitan area that is significantly warmer than its surrounding rural areas. Cities, especially large ones, have unique microclimates caused by the large amount of pavement, tall buildings that change airflow patterns, and increased concentration of emissions from cars, trucks, industries, and so on. The warmer air leads to an increase in the rate of smog-producing photochemical reactions and, along with an increased need for cooling for comfort and refrigeration, leads to greater air pollution problems. Some causes include the following:

1. Pavement, concrete, and asphalt covering the ground absorbs heat during the day and then radiates that heat back into the air at night. Often these materials are dark in color, which further increases the heat they absorb and radiate.

2. Less shade from trees and vegetation leads to more sunlight shining directly on the ground, further increasing absorbed heat.

3. The natural water cycle is disrupted. More pavement means more rainwater will run off into waterways and less will be absorbed into the ground to percolate through the soil and increase humidity. Lack of trees and other vegetation reduces transpiration, also reducing humidity and increasing temperatures.

V. **The Clean Air Act**

A. Beginning in 1955, it was clear that some type of national legislation was needed to reduce air pollution. Regulations were passed, which, in 1963, became the Clean Air Act. This legislation has been modified several times throughout the years. The provisions of the act include the following:

1. Authorization and funding studies of air quality to learn more about the presence and effects of atmospheric pollutants.

2. Setting enforceable regulations to limit emissions from stationary sources like factories and power plants as well as mobile sources like cars, trucks, and ships.

3. Developing programs to monitor and reduce acid deposition and the primary pollutants that cause its formation.

4. Establishing a program to phase out the use of chemicals that deplete stratospheric ozone.

5. Establishing a cap and trade program for SO_2.

   i.   Cap and trade is a system in which maximum allowable emissions are set for each industry.

   ii. Businesses that can reduce their emissions below the standards are awarded credits that they can sell to other businesses that cannot meet the limits.

   iii. This makes economic sense because it creates a huge financial incentive for industries to reduce emissions quickly. Allowing the sale of emissions credits means that many groups can rapidly recover the costs associated with reducing emissions.

VI. Changes in Earth's Stratospheric Ozone

A. *Ozone* in the upper atmosphere, or stratosphere, is essential to life as we know it. In the 1970s, scientists first detected a reduction in stratospheric ozone. This eventually precipitated a series of meetings resulting in legislation to phase out ozone-depleting chemicals worldwide.

B. "Good" ozone and "bad" ozone. As you read earlier in this chapter, ozone in the troposphere is a powerful pollutant. In the stratosphere, however, it is an essential blocker of excess ultraviolet (UV) radiation.

C. Chemicals that deplete ozone. Compounds that break down to release chlorine, bromine, or fluorine are the major contributors to ozone depletion. Many of these compounds are used as refrigerants, coolants, aerosol propellants, and industrial solvents.

1. Chlorofluorocarbons (CFCs or CCl_2F_2) are also greenhouse gases. In the 1970s, Roland and Molina's research showed the negative effects of CFCs on stratospheric ozone.

2. Halocarbons and halons are a type of CFC used as fire retardants, pesticides, and foam insulation.

D. Health and environmental effects. Excess UV radiation in the lower atmosphere causes great damage to the cells of living organisms. Aside from the increasing incidence of sunburn, UV radiation also greatly increases the human risk of skin cancer. Plant photosynthetic productivity is also greatly reduced due to cellular damage. Animals such as fish, amphibians, and other organisms lacking protective fur or scaly skin are also harmed directly.

E. The Montreal Protocol. One of the greatest environmental success stories to date, the Montreal Protocol is proof that global problems can be solved by a combination of international cooperation and scientific advances. The provisions of the Montreal Protocol require participating nations to phase out the use of ozone-depleting chemicals in favor of less harmful alternatives.

Test Tip

Beware! *Many students confuse the issues of ozone depletion and global climate change. This is further complicated by the fact that many ozone-depleting chemicals are also greenhouse gases. Make sure you separate the two issues in your mind. Reduction in stratospheric ozone does* not *increase global temperatures. Ozone depletion leads to increased UV penetration, resulting in increased rates of cancer for humans and loss of photosynthetic productivity in plants.*

How Human Activities Affect Earth's Water

Chapter

20

I. Major Sources of Water Pollution

A. Point sources. As in atmospheric pollution, point sources are easily identifiable, localized sources of pollutants—the drainpipe from a factory, a landfill, a sewage treatment plant, or a fish farm are all examples of point sources of water pollution.

B. Non–point sources. Non–point sources are more difficult to identify and, in the case of water pollution, currently the greatest problem.

    1. The major source of non–point pollution is *runoff*. Agricultural runoff may contain animal wastes, excess fertilizer, and toxic chemicals used as herbicides and pesticides.

    2. On a smaller scale, runoff from neighborhood landscaping may also contain many of these pollutants. Roads and parking lots often lead to runoff contaminated by motor oil, antifreeze, and other chemicals used on and in cars.

II. Water Quality Index (WQI)

A. A series of tests are used to identify the quality of a sample of water. The water quality index (WQI) was developed as a tool to measure the parameters most likely to affect the health of aquatic ecosystems. Once these measurements are made, a score is calculated that may be used in management decisions.

    Note: ppm = parts per million, and 1 ppm = 1 mg/L.

Parameter	Description/ How It Is Tested	Desired Results	Environmental Significance
Dissolved oxygen (DO)	The amount of oxygen dissolved in a sample of water. Tested with a variety of methods, including Winkler titration and digital probes. *Note:* The temperature of the water is important. Cold water has a higher capacity to hold DO than does warmer water.	High DO readings, usually in the range of 7–9 ppm or above, are best. Few fish can survive in water of less than 3 DO ppm. Low DO levels put aquatic organisms at risk of hypoxia (lack of oxygen).	All aerobic organisms require oxygen for cellular respiration. Aquatic organisms rely on oxygen dissolved in the water to satisfy this need. DO is higher in flowing water due to interaction between the surface water and oxygen dissolved in atmospheric air.
Biological oxygen demand (BOD)	Samples of water are taken in clear bottles. One is darkened to prevent sunlight from entering; one is left open to light penetration. After a set period of time, the difference in dissolved oxygen in the two bottles is used to determine the BOD.	Lower BOD. ratings are best. Most healthy aquatic ecosystems have a BOD of less than 10 ppm.	Closely related to DO testing, BOD indicates the levels of aerobic bacterial activity in a water sample. High BOD signals that there is a source of organic material to feed decomposing bacteria and encourage their population growth.
pH	A measure of the acidity or alkalinity of water. Using pH indicator paper, pH solution, or a digital probe.	Most aquatic organisms prefer a pH near neutral (7). The pH scale is logarithmic, meaning that a pH of 6 is 10 times more acidic than a pH of 7. (A pH of 5 is 100 times more acidic than 7!)	Low pH may indicate acid deposition and have harmful effects on aquatic life.
Hardness/ alkalinity	Alkalinity is a measurement of the buffering capacity of water; it is the ability of a body of water to neutralize acids without changing the pH of the water. Hardness is a measurement of the concentration of metal ions like Ca^+ and Mg^+.	Higher alkalinity is good in aquatic ecosystems; it acts as a buffer against pH drops from acid deposition. Zero to 60 ppm is soft water; above 60 ppm is in the hard-water range.	Alkalinity is associated with hard water because the calcium and magnesium ions that make water "hard" are great acid buffers. Hard water is mostly a problem in domestic use. It can leave behind cloudy deposits on glassware and requires more detergents to wash clothes or dishes.

(continued)

Parameter	Description/ How It Is Tested	Desired Results	Environmental Significance
Turbidity/ total suspended solids (TSS)	In the field, a Secci disc is lowered into the water until it can no longer be clearly seen. In the lab, a spectrometer is used.	Low turbidity is best.	Turbid water has low visibility. As turbidity increases, the euphotic zone decreases because sunlight is blocked from penetrating deeper water by the suspended particles in the water.
Fecal coliform	A sample of the water is placed in a culture dish and incubated to see if any fecal coliform bacteria colonies grow.	Zero colonies Presence of fecal coliform in a water sample is an indication that the water is likely to be contaminated by mammal or bird feces.	Although these bacteria are not harmful by themselves, they are indicators of the presence of feces, which may carry a host of pathogens that *are* harmful to human health, such as viruses, parasites, and harmful bacteria.

III. The Process of Eutrophication

A. Eutrophication

1. The literal translation of *eutrophication* from Latin is "good feeding."

2. When eutrophication happens naturally over a long period of time by gradual accumulations of sediment and other organic materials in an aquatic ecosystem, it leads to well-developed, biodiverse ecosystems.

3. When eutrophication happens rapidly, as a result of human activities, we find that "too much of a good thing" can be extremely harmful.

B. Oligotrophic Lakes

1. When a new lake is formed, it is often just a depression in the ground that fills with water through rainfall or flooding of an existing body of water.

2. In most cases, the new lake is little more than bare rock covered with water, and with few nutrients, it has few organisms.

3. These new, low-nutrient lakes are called *oligotrophic*.

C. Natural Eutrophication

1. In a natural system, nutrient-rich sediments containing nitrogen and phosphorus build up over time on the bottom and around the edges of an oligotrophic lake, gradually making more plant life a possibility. These sediments ultimately increase the natural biodiversity of the ecosystem as more nutrients make more life possible.

2. In most ecosystems, nitrogen and phosphorus are the primary limiting factors because these two chemicals are essential for building the molecules of life, such as DNA, RNA, amino acids, and phospholipids.

3. When present in low concentrations, nitrogen and phosphorus stimulate the growth of plants and, as a result, other aquatic organisms that rely on plants for food.

D. Cultural Eutrophication

1. When the accumulation of nutrients happens rapidly as a result of anthropogenic (human-caused) activities, the ecosystem is usually harmed. This is called *cultural eutrophication*.

2. Runoff. Agricultural or other nutrient-rich runoff enters a body of water. An excess of nitrogen and phosphorus can lead to algae blooms. Both nutrients are commonly found in runoff from fertilizers used in agriculture and residential landscaping, as well as sewage effluent. Because these nutrients are limiting factors, adding them into the system in excess stimulates rapid growth of plant life, especially algae.

3. Algae blooms. Algae blooms result from excess nutrients. Algae are among the most rapidly growing and reproducing organisms in aquatic ecosystems, so they usually go through a population boom in the presence of excess nutrients.

4. The euphotic zone. The *euphotic zone* in an aquatic ecosystem is the depth to which enough sunlight can penetrate to make photosynthesis possible.

    i.   If a thick layer of algae is covering the surface of the water, the euphotic zone is reduced or eliminated because sunlight is unable to penetrate to reach plants beneath the surface. Photosynthesis is impossible and plants die.

    ii. The algae on the surface are undergoing photosynthesis, but the oxygen produced is being released into the air at the surface, not into the water.

5. Decomposing bacteria consume dead plant material. Decomposing bacteria also experience a boom in population as a newly abundant source of food is available.

    i.   Because they are largely aerobic bacteria, they consume oxygen (O_2) and release carbon dioxide (CO_2).

    ii. Larger bacteria populations lead to a reduction in dissolved oxygen (DO) in the water, which is compounded by the fact that the underwater photosynthetic plants that were releasing O_2 and consuming CO_2 were killed off by the algae bloom.

6. Fish kills. Larger aerobic organisms in the system (fish) suffer from hypoxia (oxygen depletion). As DO levels continue to drop due to algae blooms, aquatic plants die off and the result is a boom of decomposing bacteria. At this point the water is visibly unhealthy, usually thick and green tinted. Often there is a strong odor of decay.

E. How to Reverse the Effects of Cultural Eutrophication

   1. Herbicides/algaecides to kill the algae blooms. Chemicals may be added to the water to kill the algae, but unless measures are taken to remove excess nutrients from the water, this is usually only a temporary solution.

   2. Pumping oxygen into the water. As a quick fix to combat low DO while more permanent steps are being taken, oxygen gas can be pumped down into the deeper depths of the water to boost DO levels so fish and other aerobic organisms avoid hypoxia. Again, this is not a permanent solution.

3. Dredging lakes to remove accumulated detritus. The build-up of decaying plant material is a major contributor to the drop in DO as decomposing bacteria populations grow in response to this drastic increase in food supply. By removing the detritus, bacterial populations can be kept in check. This is one step toward a permanent solution.

4. Bioremediation to remove excess nutrients. The best way to avoid cultural eutrophication is to bring water nutrients to normal levels. First, the source of the runoff or effluent must be located and stopped. Next, certain species of fast-growing plants (duckweed) can be used in conjunction with some or all of the methods mentioned above to pull excess nutrients out of the water. Duckweed is convenient because it is a commonly found aquatic plant (not likely to be an exotic) that floats on top of the water and can be scooped out once it has done its work. Any remaining duckweed is kept in check by lower nutrient levels in the remediated water.

Make sure you know and understand the entire process of eutrophication. This is an extremely important concept that shows up on every exam, usually relating to more than one question. Understanding how all the factors (DO, algae blooms, and fish kills) are connected is the key.

IV. Municipal Wastewater

A. Definition of Municipal Wastewater

1. Water discharged by domestic residences, businesses, industry, and stormwater all constitute wastewater.

2. Wastewater is treated in wastewater treatment plants, then either re-used or returned to nature.

B. The Steps of Wastewater Treatment

1. Primary treatment = mechanical filtration.

   i. Water is passed through a series of screens to catch large debris and undissolved materials.

2. Secondary treatment = biological filtration

    i. Water is then moved to large aerated ponds, and decomposing bacteria in activated sludge are added to digest away any dissolved organic matter.

    ii. After exposure to the bacteria, water is treated with disinfecting chemicals, like chlorine or ozone, to kill any living organisms present.

3. Tertiary treatment = further purification

    i. Nutrient removal. Using advanced filters or even treatment wetlands, excess nutrients like nitrates and phosphates are removed from treated wastewater.

    ii. Advanced disinfection. Some wastewater treatment plants use strong ultraviolet (UV) lights or reverse osmosis as additional disinfecting steps.

4. Treated wastewater is either returned to the environment by being released into a local waterway or, in some large cities with water shortages, it may be further treated for purification and returned into the municipal water supply to be reused in households, industry, and public works.

C. Is Treated Wastewater "Clean"?

1. Currently, treated wastewater is considered "clean" when it tests within acceptable levels for nitrates, phosphates, bacteria, and disinfecting chemicals like chlorine.

2. Other potentially worrisome chemicals, like pharmaceuticals, including antibiotics and human hormones used in birth control pills, are not removed or tested for.

V. Other Water Quality Concerns

A. Heavy Metal Contamination

1. Heavy metals are extremely toxic to humans at relatively small concentrations.

    i. In aquatic ecosystems, heavy metals tend to accumulate in large predatory fish thanks to biomagnification and, as a result, people should limit their intake of these fish in their diet.

   ii. The neurotoxic effects of heavy metals are not limited to humans. Aquatic organisms also suffer a wide array of health problems in water with elevated levels of metals such as lead, mercury, zinc, and arsenic.

2. Mercury and, to a lesser extent, arsenic are products of the combustion of coal to produce energy. They end up in aquatic ecosystems as small airborne particles that are carried in the wind and eventually settle to the ground or in bodies of water either as dry deposition or mixed with rainwater.

3. In addition to being directly harmful to organisms, acidification of water actually makes heavy metals in sediments more soluble, thereby increasing concentrations of zinc, mercury, and other heavy metals in the acidified water. This drop in pH may be caused by acid deposition or also by acid drainage from coal-mining operations.

B. Oil Spills

1. Disasters such as the wreck of the Exxon *Valdez* in 1989 and more recently the BP Deepwater Horizon explosion and leak in 2010 show that drilling for and transporting oil has the potential to do great environmental harm when accidents happen.

   i. Common methods of cleaning up oil spills include mechanical containment, chemical methods, and biological agents.

➤ Mechanical containment. Large booms filled with absorbent materials (even human hair!), skimmers, and other types of barriers are placed in the water to attempt to contain the spread of oil.

➤ Chemical methods. Detergents and other dispersing agents attempt to break down oils, while clumping or gelling agents cause the oil to form large globules that can be more easily scooped or skimmed from the water.

➤ Biological agents. Naturally occurring oil-consuming bacteria will, over time, eventually break down the oil. To dramatically hasten the process, water may be fertilized with nutrients such as nitrogen and phosphorus to encourage bacteria growth. Water may also be seeded with oil-consuming bacteria to boost populations.

VI. Pollution Case Studies

A. Cuyahoga River Fire

　1. In 1969, the Cuyahoga River in Cleveland, Ohio, was so polluted with industrial wastes and oils that it caught fire. Although the fire lasted only a short time and was not the first, photos were taken by reporters from national magazines and newspapers and the fire became a news sensation. After this, Cleveland became notorious for its problems with industrial pollution, and national pressure increased and led to legislation such as the Clean Water Act and contributed to the development of the Environmental Protection Agency (EPA).

B. Lake Erie

　1. Lake Erie is the shallowest and warmest of the Great Lakes in the United States. Surrounded by urban development, it was subject to mass cultural eutrophication in the 1960s from excess nutrients flowing into the lake from surrounding industry and households. The primary culprit was phosphorus, which was largely used in detergents. Phosphorus is usually in scarce supply in aquatic ecosystems and the primary limiting factor for plant growth. When excess phosphorus was added to Lake Erie, massive algae blooms caused the death of aquatic plants and fish from the drastic reduction in dissolved oxygen in the lake water.

VII. Water Quality Legislation

A. Clean Water Act (CWA)

　1. Originally passed in 1972 and revised several times since, the mission of the Clean Water Act is to restore and maintain

the chemical, physical, and biological integrity of America's waterways to support wildlife and recreational activities. The provisions of the CWA are intended to reduce and prevent both point and non–point sources of water pollution in the nation's surface waters. There are no provisions specific to groundwater protection.

B. Safe Drinking Water Act (SDWA) of 1974

  1. The focus of the Safe Drinking Water Act is to maintain the purity of any water source that may potentially be used as a source of drinking water. This includes both surface water and groundwater.

Legislation questions don't come up very often on the AP Environmental Science exam, but when they do, they are commonly in the free-response section of the exam. If you are asked to discuss a piece of legislation relating to water, you are safe to discuss either the Clean Water Act or the Safe Water Drinking Act. If wildlife or aquatic ecosystems are involved, choose the CWA. If human health is a focus of the question, then the SWDA is a better bet. The provisions of both acts are very broad and cover a variety of circumstances.

REFERENCES

USGS Water Quality Information: Hardness and Alkalinity. *http:// water.usgs.gov/owq/hardness-alkalinity.html*

Alkalinity and Stream Water Quality. *http://www.water-research. net/Watershed/alkalinity.htm*

Biological Agents/Emergency Management/U.S. EPA. *http:// www.epa.gov/emergencies/content/learning/bioagnts.htm*

Solid and Hazardous Waste

 Municipal Solid Waste

A. Any trash or garbage that is produced in households, schools, and businesses and is disposed of through local waste collection is considered *municipal solid waste*. Here are some important facts to know about solid waste.

B. Paper is the No. 1 type of solid waste produced (by weight) in the United States. More paper is thrown away than any other type of material.

C. Options for the Disposal of Solid Waste

   1. Landfills

      i. Landfills are the most common disposal method in the United States for solid waste. In its simplest form, a landfill is any place that is used to store trash and garbage.

      ii. Sanitary landfills are designed to isolate waste from the environment until it is deemed sufficiently decayed to no longer be potentially harmful.

      iii. Sanitary landfill sites are chosen based on geology. An important consideration is soil type. High clay content in soil is preferred because it acts as a barrier to chemicals that may leak from the landfill. These chemicals are called *leachate*.

      iv. One major concern with landfills as a disposal method is the possibility of leachate seeping into the groundwater and contaminating potential drinking water sources.

v. Waste is often crushed or compacted before being placed in landfills. Crushing trash often creates smaller particles of waste with a higher surface area. This is a problem because it increases the potential for leaching.

2. Incineration

i. *Incineration* is the burning of waste.

ii. While incineration is great at reducing the volume of waste that must be stored, the major drawback is that combustion releases toxic chemicals into the atmosphere.

3. Composting

i. Any biodegradable material is a candidate for composting. A material is considered biodegradable if exposure to natural elements, such as water, sunlight, and bacterial decomposition, can break it down. Composting can be accomplished on a large or small scale.

ii. Individual homeowners often choose to do backyard composting by collecting kitchen scraps and yard waste in a pile and ultimately using the finished compost as a rich organic fertilizer.

iii. Composting can also be industrially scaled. More and more municipalities are using large, pressurized digesters to rapidly break down organic wastes. The resulting compost may then be used as a layering agent in landfills or spread as fertilizer.

4. Recycling

i. *Recycling* is the collection and recovery of useful materials that can be remade into new products.

ii. The Pros of Recycling

➤ Recycling keeps many types of waste out of landfills.

➤ Materials like glass and most types of metal can be recycled many times without much loss in the purity or usefulness of the materials.

➤ Recycling reduces the need for environmentally damaging mining and extraction of metals like bauxite (for aluminum) and copper.

➤ Recycling paper reduces deforestation for wood pulp.

➤ When compared to the energy required to extract, purify, and process many raw materials, especially metal ores, recycling is much more efficient and, in many cases, produces fewer pollutants.

iii. The Cons of Recycling

➤ Recycling is energy- and often water-intensive. In areas where water is limited, processes like recycling paper may be difficult to justify.

➤ In most cases, recycling a material means that it must be melted and re-formed into a new product. In addition to requiring large inputs of energy, the melting process may release toxic gases into the atmosphere or result in harmful chemical by-products that must be disposed of.

➤ Recycling can be expensive. Consider the costs of providing recycling bins to every household, maintaining and powering the collection vehicles, paying workers to collect and process materials, and maintaining the recycling facility. If there is not a high demand for the resulting recycled products, municipalities often must use tax dollars to support the process.

➤ *Downcycling* is when the products made from recycled materials are of a lesser quality or purity than the original. Some materials, most notably plastics and paper, are more prone to downcycling.

— For example, used paper often contains inks and dyes used in the printing process. It may be impossible to remove all of the impurities when making new paper, so writing paper and newsprint is often downcycled into products like cardboard or paperboard. Because plastics come in so many varieties, they are often mixed together when melted, so the resulting re-formed plastic is a combination that is of a lower grade than the original(s).

5. Integrated Waste Management

   i. Most municipalities use a combination of disposal methods to manage their waste.

   ii. Where appropriate, recycling is encouraged or enforced. Some materials are incinerated or composted and the remainder is placed in a sanitary landfill.

6. Reducing and Reusing

   i. We all know the three R's: reduce, reuse, and recycle. Reducing and reusing help to keep waste from being generated in the first place. In terms of environmental sustainability, reducing and reusing are the best options of all.

   ii. *Reducing* simply means to throw away less.

      ➤ Buy products with less packaging. Bring reusable grocery bags to the store instead of using the disposable plastic ones. Avoid buying and using paper plates, napkins, and any disposable products. Use a metal water bottle that you fill from the tap instead of buying bottled water.

      ➤ Reducing helps in several ways: Landfill space is saved by the reduction in trash, fewer materials need to be mined or extracted from the environment, and the consumer saves money that would otherwise be spent on disposable products.

   iii. *Reuse* means to think of another way to use the item.

      ➤ Instead of throwing something away when it is no longer useful, think of other ways it could be put to use. Donate unwanted items to thrift shops or shelters. Use "trash" for art projects: egg cartons make great paint trays; broken dishes are good for mosaics; cardboard boxes can be made into a child's playhouse, and so on.

II. Hazardous Waste

A. *Hazardous waste* is a category of waste that includes materials that are known to be harmful to humans and other living organisms.

  i. Hazardous wastes contain chemicals that are poisonous, toxic, or otherwise known to reduce human health and/or lifespan through exposure.

  ii. Hazardous wastes may not be disposed of in regular landfills. They must meet specific requirements to endure safe disposal.

B. Criteria for Defining Hazardous Waste

  1. Negative Health Effects

  i. Materials known to cause a specific health effect can be carcinogens, teratogens, or mutagens. Carcinogens cause cancer, teratogens cause birth defects, and mutagens increase genetic mutations, and all are classified as hazardous.

  ii. This class includes heavy metals (mercury, lead, cadmium, and others), pesticides, many household and industrial chemicals, and nuclear wastes.

  2. Ignitability

  i. Gasoline, solvents, and some paints are examples of materials that easily catch fire and thus would be dangerous in a mixed waste landfill.

  3. Explosiveness

  i. Many chemicals are reactive enough to explode or release toxic fumes if they are not properly stored or disposed of.

  4. Corrosiveness

  i. Substances like strong acids or cleaning agents may be so corrosive that they can eat through metal barrels or containers.

C. Methods for Neutralizing or Disposing of Hazardous Wastes

1. Biological treatment. A variety of bacterial decomposers may be used to break down hazardous organic wastes or other biological materials.

2. Chemical treatment. In some cases, chemical reactions are used to neutralize chemical wastes, thus creating products that are far less toxic.

3. Burial. Materials like radioactive wastes are contained in metal drums and then placed for long-term storage in underground, cement-lined facilities because there is no safe way to break them down.

D. Legislation About Hazardous Waste (CERCLA and RCRA)

1. The Comprehensive Environmental Response, Compensation, and Reliability Act (CERCLA)

   i. Also known as the Superfund Act, CERCLA was enacted to handle industrial contamination in sites where no direct individual or party could be held responsible for cleanup.

      ➤ For example, a company released arsenic into the soil many years ago and then went out of business. The site was sold to a new owner. Who is now responsible for cleaning up the arsenic?

2. The Resource Conservation and Recovery Act (RCRA)

   i. Commonly called the "cradle to grave" act, RCRA sets specific regulations concerning the manufacture, transport, storage, use, and ultimate disposal for a host of hazardous chemicals. Its major provision requires extensive documentation at every step to ensure that hazardous wastes are disposed of properly.

Test Tip

Don't bother learning the full names of either of these acts for the exam. They are always referred to by their acronyms—CERCLA and RCRA.

Changes in Earth's Climate

I. The Greenhouse Effect

A. The *greenhouse effect* occurs when gases in the lower atmosphere (troposphere) trap infrared radiation from the sun. This heat radiates back to Earth and further warms it.

B. The major greenhouse gases are listed in the following table.

Greenhouse Gas	GWP*	Major Sources	Remediation or Reduction
Water vapor (H_2O)	2	Nonanthropogenic (natural). Warming naturally increases atmospheric water vapor due to evaporation.	Not human-made, so there is little we can do to control it.
Carbon dioxide (CO_2)	1	Burning fossil fuels.	Move energy generation from fossil fuels to other sources such as hydro, wind, solar, nuclear. Carbon sequestration.
Methane (CH_4)	20	Livestock. Anaerobic respiration from decomposition in swamps and landfills.	Capture, compress, and burn as natural gas. CH_4 degrades quickly in the air so reducing emissions would have an almost immediate effect.
Nitrogen dioxide (N_2O)	300	High-temperature combustion, especially in automobiles.	Electric or hybrid cars.

(continued)

Greenhouse Gas	GWP*	Major Sources	Remediation or Reduction
Chlorofluoro-carbons (CFCs) (CCL_2F_2)	6500	Not usually found in nature; purely anthropogenic. Refrigerants.	Montreal Protocol banned CFCs. CFCs persist in the atmosphere for hundreds of years, so even though they are no longer being released, they will continue to affect the atmosphere.
Ozone (tropo-spheric) (O_3)	Potent, but no specific number	A component of photochemical smog as a result of automobile emissions, burning biomass, and industrial pollution.	Reduce emissions of smog-causing gases (NO_x, SO_2) and particulate matter by using catalytic converters on cars and smokestack scrubbers.

*GWP = Global warming potential in comparison to carbon dioxide. These numbers are approximate. The greater the number, the higher the potential to trap heat in the atmosphere.

Source for GWP: National Climate Data Center (http://www.ncdc.noaa.gov/oa/climate/globalwarming.html)

Every past AP Environmental Science exam has addressed, in some way, the fact that the burning of fossil fuels and biomass are the major sources of increased carbon dioxide in the atmosphere.

II. Focus on Carbon

A. Fossil Fuels

    i. All life on Earth contains carbon.

    ii. Fossil fuels originate from carbon compounds naturally released from living things as they slowly decay over long periods of time and intense pressure.

      ➤ Petroleum (oil) comes from the bodies of prehistoric plants and animals. Coal comes from the plants in and around ancient swamps.

B. Causes of Seasonal Fluctuations of Carbon

➤ Every year there is a measurable cycle of carbon dioxide caused by seasonal changes in photosynthesis occurring in the world's forests. This cycle is represented by the dashed line on the graph. The dotted line represents the mean average for each year.

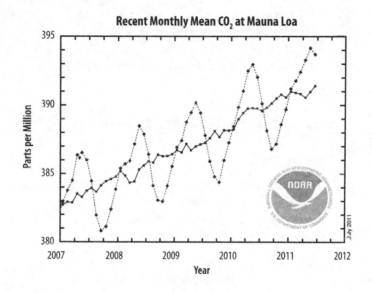

Recent Monthly Mean CO_2 at Mauna Loa

➤ The process of photosynthesis allows plants to take in carbon dioxide and, in the presence of sunlight, manufacture glucose and oxygen. Therefore, more photosynthesis lowers atmospheric CO_2.

➤ In the cold winter months (or dry season near the equator), trees in deciduous forests lose their leaves, and evergreens go into a dormant state to conserve water. This leads to a significant, worldwide drop in photosynthesis during winter in the northern hemisphere because there is much more land and vegetation north of the equator. As a result, atmospheric CO_2 goes up.

➤ For the same reason, there is a measurable, worldwide decrease in atmospheric CO_2 during the warmer months (or wet season) in the northern hemisphere as photosynthesis reaches its peak.

C. Burning Fossil Fuels

1. Burning fossil fuels increases atmospheric carbon by releasing carbon that had been sequestered underground in coal and petroleum.

D. Clear-Cutting

1. Clear-cutting a tropical rainforest increases anthropogenic climate change because of a reduction in CO_2 uptake by plants (from photosynthesis), and burning the logged materials releases carbon into the atmosphere.

III. Evidence for Global Climate Change

A. Ice Core Data

1. Data from Antarctic ice cores can be used to measure atmospheric gas composition from hundreds of thousands of years ago. Ice core data shows a distinct correlation between atmospheric CO_2 concentration and air temperature. See the following graph.

The Vostok (Antarctica) Ice Core Record
Carbon Dioxide versus Temperature for the last 420,000 years

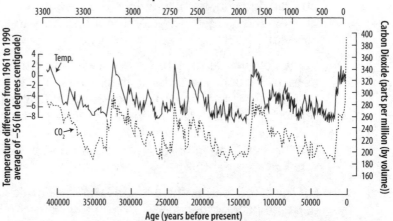

2. For the majority of the graph, CO_2 and temperature follow a very similar pattern.

3. Only at the very far right (near the present time) do we see a discrepancy: CO_2 concentration is sharply increasing to ever-higher levels than anywhere else on the graph (nearly twice as high).

4. During this same time, temperatures are at the high end of the cycle. The current concern is that, because CO_2 is a greenhouse gas, the rise in CO_2 is likely to lead to a further rise in temperature that will be higher than any temperature recorded on Earth in the last 400,000 years.

5. However, correlation does not equal causation. Scientists cannot be absolutely sure if increased CO_2 is causing the rise in temperature. It is possible that the rise in temperature may be causing the rise in CO_2, or perhaps there is some other factor that hasn't yet been discovered. However, because CO_2 is a known greenhouse gas, there is a very high likelihood that a significant rise in temperature will result from a spike in atmospheric CO_2 concentration.

B. Historic Data

1. Historic data show that global temperatures have increased by 0.5°C in the last 100 years.

2. The data also show that the rate of increase for temperature continues to increase.

3. It is expected that within the next 100 years, global temperatures could increase by as much as 4°C.

C. Effects of Climate Change on Earth's Ecosystems and the People Who Rely on Them

1. Changes in Ocean Levels. As the atmosphere and oceans warm, a significant rise in the overall sea level will be seen.

   i. The major cause of sea-level rise is thermal expansion of oceans. As liquids warm, the molecules making them up are energized and thus experience more energetic collisions, causing them to spread out. This leads to a slight overall increase in volume that may not be noticeable in a small pan of water, but in an area as large as the Earth's oceans, this will be one of the major causes of sea-level rise.

ii. Melting ice is another significant cause of sea-level rise.

   ➤ This is true in the case of land-based ice such as mountain glaciers, the Greenland Ice Shelf, and Antarctic continental ice. As this ice melts, the water runs from land to sea, resulting in more ocean water and a sea-level rise.

   ➤ This is *not* true in the case of sea-based ice. Sea-based ice such as icebergs and most of the Arctic ice do not affect sea level because they are already in the water. As they melt, they just become part of the water in which they were floating (like ice cubes in a glass of water—the glass does not overflow as the ice cubes melt).

iii. Small increases in sea level will cause flooding of coasts and low-lying areas. In estuaries and coastal ecosystems, this will lead to the loss of wetlands, marshes, and intertidal zones that may be important breeding or shelter grounds for birds, fish, and shellfish. This could cause the decline or total loss of these species.

iv. Roughly 40 percent of the world's population lives within 60 miles of a coastline. Rising water will severely affect coastal communities because of the following factors:

   ➤ Loss of property and property values as homes flood and land erodes from increased storm surges and loss of beach sand or mangrove stands.

   ➤ Estuary flooding, which could lead to the loss of fish and shellfish species. Humans who rely on them as a source of income or food will suffer.

➤ Loss of beach and fishing areas may lead to collapse of tourism in coastal areas—which is often their major source of income and jobs.

➤ Intrusion of saltwater into water supplies for drinking and irrigation can leave communities with no source of freshwater.

➤ Increased conflict because people displaced by flooding will migrate inland into already crowded areas.

2. Large River Systems. Large river systems like the Colorado River in the United States or the Yangtze River in China could experience the following:

   i. Warmer temperatures will cause increased spring melt of mountain snow pack. For a time this will increase the flow of water in the river (which is good because many people rely on these large rivers for drinking water, hydroelectric power, etc.). In the long term, however, the increased spring melt reduces the overall size of the mountain glaciers, and eventually there will be no snow pack left to supply the river.

   ii. Depending on where the river is located, global climate change will likely mean either increased or decreased precipitation. Increased precipitation would increase the input to surface water and groundwater (good), while runoff from the rain falling on land and running into the river would lead to increased erosion and sedimentation (bad).

   ➤ Decreased precipitation would do just the opposite—reduce input to surface and groundwater (bad) and decrease erosion and sedimentation (good)

   iii. There is also projected to be an increase in the frequency and severity of storms. Stronger storms would lead to increased sedimentation and flooding, and greater volumes of runoff.

3. Biome Loss. As a result of global climate change, many biomes will no longer exist where they are currently found.

   i.   Biomes are determined by temperature and precipitation. As global patterns of both change, many of the warmer biomes will expand outward toward the poles. This means that the tropics could expand into the southern United States, while the temperate zones will shift northward into the region that is now boreal forests.

   ii. Earth's atmosphere will contain more water vapor as surface water is warmed and evaporates. This will cause changes in the distribution and frequency of precipitation that are difficult to predict.

   iii. Specialist species with a narrow range of tolerances are likely to be driven to extinction when their fragile ecosystems change too rapidly for them to adapt.

   iv. The distribution of many insect species will grow as warmer climates and increased rainfall means more ideal breeding conditions and less winter die-off with the decrease in cold winter months. Insect vectors transmit many infectious diseases, such as malaria and West Nile virus. Expanded distributions of mosquitoes and other insect vectors will allow these diseases to spread to currently unaffected areas. As insect populations expand, there is also more potential for crop damage from increased pest attacks.

D. The albedo effect is an example of a positive feedback loop that works to accelerate global warming, especially at Earth's poles.

   1. Because ice is reflective and light in color, it has high albedo and reflects most of the sun's radiation (heat) away from Earth's surface. Warmer temperatures cause ice to melt, allowing more sunlight to fall on the underlying land or water instead.

2. Land and water are both darker in color and less reflective than ice, so more of the heat is absorbed and less is reflected (low albedo). This leads to an increase in temperature, which in turn accelerates the melting of ice.

3. This is a positive feedback loop because the increased temperature creates a situation that leads to greater and greater increases in temperature. This is often referred to as a vicious cycle because once it begins, it is difficult to halt or reverse.

IV. Efforts to Reverse the Warming Trend—Laws and Treaties

A. The Kyoto Protocol

    i. The Kyoto Protocol was developed at a world summit in 1997 to address reduction of greenhouse gas emissions. Many promises were made by many nations, but no real progress was made. The United States announced that it would not sign, and soon after the whole movement lost support.

B. Conferences on Global Climate Change

    i. There have been several meetings since, most notably the Copenhagen Climate Summit in 2009. Again no real binding resolutions were ratified.

This legislation can be confusing because of the controversy surrounding global climate change and the fact that these treaties have never been fully ratified. There has never been a question on the AP Environmental Science exams about this legislation, probably because it is so confusing. If you recognize that the Kyoto Protocol is connected to global warming and know that the only two major countries who refused to sign the protocol are the United States and Australia (who happen to be two of the greatest emitters of greenhouse gases) you should be all set.

Math Practice

1. A typical family in Florida uses 2,000 kwh per month of electricity during the hot summer months. All of their electricity is generated by their town's coal-fired power plant. This power plant emits 100 kg of CO_2 per MMBTU (million BTUs) generated at the plant. For every kwh of electricity delivered to a home, 3,412 BTUs of energy are generated at the plant. Use this information to determine how many Kg of CO_2 are emitted per month to generate electricity for this family.

2. Look back at the graph titled "Recent Monthly Mean CO_2 at Mauna Loa." Using the graph, determine the percentage increase in CO_2 from 2006 to 2010.

ANSWERS:

1. $$\frac{? \text{ kg CO}_2}{\text{month}} = \frac{100 \text{ kg CO}_2}{\text{MMBTU}} \times \frac{\text{MMBTU}}{1 \times 10^6 \text{ BTU}} \times \frac{3412 \text{ BTU}}{\text{kWh}}$$
$$\times \frac{2000 \text{ kWh}}{\text{month}} = \frac{682.4 \text{ kg CO}_2}{\text{month}}$$

2. $$\frac{388 - 381}{381} = \frac{7}{381} = 0.018$$
$$0.018 \times 100\% = 1.8\%$$

Test Tip

Based on an analysis of past AP Environmental Science exams, the most important things from this section to remember are:

• Burning fossil fuels is the major source of atmospheric carbon emissions.

• The rise in ocean levels from global warming will occur primarily because of thermal expansion of ocean water and not from the melting of land-based ice. (Melting ocean ice will also not significantly change sea levels.)

REFERENCES

Seasonal fluctuations in CO_2 (Mauna Loa) graph: Dr. Pieter Tans, NOAA/ESRL:

Vostok ice core graph (from Petit et al., 1999 *Nature*). *http://pre-earth.net/vostok-temp-vs-co2.gif*

Species Extinction
and Loss of Biodiversity

 I. Biodiversity

A. *Biodiversity* is a measurement of the variety of all the different types of life on Earth.

B. Three Types of Diversity

　1. *Genetic diversity* is the variety of different genetic traits and alleles found in the genome of a species.

　　i. High genetic diversity is extremely beneficial to a species.

　　　➤ It confers resistance to epidemic diseases because there will likely be some individuals in a population with genetic resistance who are able to survive an otherwise fatal outbreak.

　　　➤ Genetic diversity is also essential in the survival of a species during a time of ecological changes because it allows a wider range of tolerance for environmental factors among individuals of the species.

　　ii. If a species is greatly reduced in numbers, a genetic bottleneck may occur.

　　　➤ If a population has a large number of individuals, each individual has its own set of genetic alleles. A population with many individuals is likely to have a larger number of possible alleles. When a population is reduced to just a few individuals, the genetic pool of that population is now made up of only the alleles found in the individuals who are left. Even when the population grows again, the gene pool of the population is reduced because all the new members of the population are offspring of the few who survived the die-off.

➤ The African cheetah is a famous example of genetic bottleneck. In some form, cheetahs have roamed the Earth for millions of years, but today they are expected to face extinction within a few generations. Every cheetah alive today is almost genetically identical to every other cheetah because, at some point in the recent past, the cheetah population was reduced to a handful of individuals. Most evidence points to the end of the last ice age about 10,000 years ago, when many other large animals were driven to extinction by natural changes in the environment. Currently, there are about 10,000 cheetahs in the wild, but all are highly inbred descendants of those original few survivors.

2. *Species diversity* is a measurement of how many different species are present in an ecosystem (species richness). The relative abundance of each species (species evenness) is also taken into account. When people use the term *biodiversity*, they are usually referring to species diversity.

3. *Ecosystem diversity* is a representation of all the different types of ecosystems present in an area. For example:

   i. Within the state of Florida, Everglades National Park is noted for high ecosystem diversity. Within the park, mangrove swamps, hardwood hammocks, cypress stands, coastal lowlands, and estuary ecosystems are found.

   ii. When land is cleared and converted to agricultural or urban use, ecosystem diversity is greatly reduced, which in turn has a negative effect on species and genetic diversity.

C. Importance of Biodiversity

   1. Interconnectedness of Life

     i.   If we remove one or more species from an ecosystem, we risk the destruction of the entire system. We rely on ecosystems to provide us with many of our material needs such as food, lumber, and clean water. If we degrade these systems, we put our precious resources at risk.

   2. Future Resources

     i.   It has been said that clearing a rainforest is like burning a library without ever reading any of its books. Scientists are constantly making new discoveries of ways to use our knowledge of other organisms to improve our lives. Many pharmaceuticals are plant-based chemicals; agronomists are using genes from rainforest plants to engineer new varieties of crop plants; and materials science engineers use our increasing knowledge of how living organisms accomplish tasks like waterproofing, climbing, and gliding to develop new technologies and materials.

   3. Aesthetic and Intrinsic Value

     i.   Many people believe that we do not need to justify preservation of biodiversity based on how it benefits humanity. The idea is that all organisms have every right to exist on Earth. Humans are just one of the millions of species on Earth, and we do not have any special rights or ownership to use and destroy other species according to our own will.

     ii.   Religious and cultural beliefs often determine our views on biodiversity. For example, Christianity holds humans as the stewards of the Earth, whereas Hinduism teaches that humans should live in harmony with nature.

     iii.   *Ecotourism* is a type of recreation in which people plan vacations specifically designed to visit a unique ecosystem or to see a specific type of plant or animal.

> ➤ Central American countries such as Costa Rica and
> Belize have found that they can make far more
> money by preserving their tropical ecosystems and
> attracting tourists for rainforest hikes, rafting trips,
> and tours than they would earn by cutting down
> the rainforest for development.

II. Endangered Species

A. A species is considered *endangered* if it is at risk of becoming
extinct in the near future. Extinction occurs when the last member of a species dies. Some notable extinct species include the
dodo, passenger pigeon, and the Carolina parakeet.

B. Characteristics That Make a Species Vulnerable to Extinction

1. Specialist species. Specialist species get their name from the
fact that they have very special requirements for survival.

   i. Feeding. The giant panda is an excellent example of a
   specialist feeder. Ninety-nine percent of its diet consists
   of bamboo. Pandas need access to bamboo forests in
   their habitat because they will not adapt to eating other
   foods. This makes them very dependent on a stable
   habitat. Changes in their habitat that reduce bamboo
   forests have left giant pandas with a shortage of food
   sources; as a result, they are critically endangered.

   ii. Symbiotic relationships. There are many instances of
   relationships between two species that have become
   highly specialized. For example, many species of plants
   have flowers that are perfectly adapted to the beak of a
   species of bird or insect that pollinates them.

2. K-strategist. Many of the life history traits that define a species as a K-strategist also put them at risk of extinction.

   i. Large body size. Large species require large amounts of
   resources including food, clean water, habitat for roaming, and so on. As a result, these species tend to have
   small population sizes when compared to r-selected
   species. When a K-selected species finds itself in compe-

tition with humans for access to these resources, there is an increased likelihood of extinction. Examples include grizzly bears, African and Indian elephants, and American bison (buffalo).

    ii.   Low birth rate. K-selected species generally take several years to reach sexual maturity, have a long gestation period, and give birth infrequently (often only once a year or less) to a small number of offspring (one to three at a time). This characteristic makes them less adaptable to rapid changes in their environment. r-selected species, like E. coli bacteria, can rapidly adapt through evolution because they have a generation time of minutes. K-selected species often have generation times of twenty or thirty years.

3. Perceived as a Pest Species

    i.   The gray wolf was driven to the brink of extinction in North America in the 1930s. Their natural prey is ungulates (hoofed mammals). Not surprisingly, they occasionally killed livestock in their hunt for food. As a result, ranchers in the central United States launched an eradication campaign in which every wolf was hunted and killed. This drove the gray wolf to extinction across much of its former range.

    ii.   In 1995, gray wolves were reintroduced into Yellowstone National Park as part of their protection under the Endangered Species Act. Today, the gray wolf population in North America is growing to the point where it is likely that they will soon be removed from the endangered species list.

4. The effects of biomagnification. Top predators are prone to the effects of biomagnification. If you remember how toxins travel through the trophic levels of a food web, you know that the organisms at the top will accumulate the highest concentration of the toxin in their tissues. Sometimes this kills them outright, but more often it will disrupt their ability to reproduce, find food, or otherwise thrive in the environment.

i. In the 1950s and 1960s, the use of the pesticide DDT caused bald eagles to accumulate DDT in their tissues as they ate fish contaminated with DDT. DDT interfered with calcium deposition in birds, which caused them to lay thin-shelled eggs. Bald eagle numbers severely declined because they were unable to lay fully calcified eggs. As a result, their eggs could not be incubated to hatching. Only since the use of DDT has been banned in the United States have bald eagle populations begun to recover fully.

5. Commercially valuable. Some species are cursed by a beautiful fur or hide, delicious flesh, or some other feature that makes them highly valuable to humans.

    i. Species like American alligators, hawksbill sea turtles, and Sumatran tigers have all been hunted for their hide, shell, or fur. These products are then made into decorative items like handbags, shoes, jewelry, or rugs.

    ii. Sperm whales, right whales, and fin whales are just a few of the species of whale that were regularly hunted beginning in the seventeenth century and continuing unregulated through the mid-twentieth century. It is estimated that before whaling was regulated in the 1970s, more that 1 million sperm whales were harvested by whalers. Sperm whales were especially prized because of a waxy substance called *spermaceti* that is derived from whale oil. This odorless and tasteless wax was used widely in making candles, cosmetics, lubricants, and medicating ointments. More recently, other chemical formulations have been developed with similar properties. Other species of whales are hunted for whale oil and whale meat. Although commercial whaling is banned in the waters surrounding the United States, other countries like Norway and Iceland continue whaling on a more limited basis.

6. Traditional medicines.

    i. There are two African species of rhinoceros, the white rhino and the black rhino. Both are critically endangered due to habitat loss and hunting. The rhinoceros is hunted for its horn, which is prized as a raw material for carving and an ingredient in Chinese traditional medicines.

    ii.   Several other animal species are aggressively hunted or *poached* (illegally hunted) due to demand as ingredients in Chinese traditional medicine. Some of these include all species of tigers (for their bones), Asiatic black bear (for its bile), musk deer (for its musk), and more than thirty-five species of seahorse (their whole body is used).

C. Causes of Extinction

  1. Anything that alters the natural environment faster than its inhabitants can adapt is a risk factor for extinction. Here are a few ways that humans cause extinction.

  2. Habitat loss is the No. 1 cause of species extinction.

    i.   As human populations continue to grow, we are encroaching more and more into formerly undeveloped areas. We are converting more land for agriculture to feed our growing population, and urbanization is increasing on a global scale as well.

    ii.   Fragmentation of habitats due to highways, power line corridors, and so on, reduces area for home ranges of large roaming animals. This in turn reduces gene flow within a species, creating genetically distinct subpopulations.

    iii.   Pollution is another cause of habitat loss. If we release pollution that acidifies a lake beyond the range of tolerance for the species that live there, we have essentially destroyed their habitat, putting them at risk for extinction.

  3. Hunting/Poaching

    i.   Humans have hunted many species to the edge of extinction. Humans continue to hunt for meat, for vanity products like combs made of hawksbill sea turtle shell, or (in the case of American bison) merely for sport. There was a time when it seemed the Earth's resources were so abundant, it never occurred to people that extinction was possible. We see now that it is not only possible, but a real threat to many species.

4. Competition from Invasive Non-Native Species

  i. An exotic species is one that has been taken from its natural habitat and transplanted to a new place. Sometimes this happens naturally through a storm event or via seed dispersal by birds and bats, but more often it is a result of human activities. Whether intentional or unintentional, the consequences can be devastating for native species.

 ii. Exotic species in a new habitat often outcompete native species with similar niches because the exotic species no longer has natural predators, diseases, or other ecological checks when it is removed from its natural system.

iii. Bilge water transport. When a fully loaded cargo ship comes into port and is unloaded, the ship rises in the water as it becomes lighter. This makes the ship unstable; before it leaves port to sail homeward, the ship will fill large tanks in its hull with ballast, or bilge water. This adds weight to the ship and lowers it in the water, increasing its stability. When it returns to its homeport, the ship releases the bilge water from its tanks prior to being reloaded with cargo. In this way, thousands of exotic species have been transported all over the world to be released in ports thousands of miles from their native habitat. The introduction of these native aquatic species has caused major problems worldwide, especially in the Great Lakes of the United States.

 iv. Biological control gone wrong. Sometimes we intentionally import a species to do a job for us. Generally, we think that using natural relationships among species (biological control) to serve our purposes is a more sustainable alternative to less natural means, and often it is. But not always.

  ➤ In 1935, cane toads were introduced into northern Australia as a method of biological control for sugar cane beetles. Cane toads had been used successfully in Puerto Rico and Hawaii, so the Australians saw this as an inexpensive and simple way to control cane beetles in their sugar cane crops without the need for pesticides. Cane toads have become one of the most aggressive invasive exotic species known. They eat anything that can fit in their mouths, breed rapidly, and produce a powerful toxin that is often lethal to any animal that tries to eat them. They do not seem to eat cane beetles at all. The cane toad has been a particular threat to many small species of marsupial, which are either eaten by the toads or killed by their toxins as a result of contact with them.

➤ During the mid-twentieth century, as south Florida was undergoing a great deal of agricultural and urban development, a huge, swampy wetland known as the Everglades was seen as a major impediment to progress. The standing water was a breeding ground for mosquitoes and other insects. There were also alligators, crocodiles, and many species of snakes. People also found that if the swampy land were dried out, the soil that had been waterlogged for years was rich with a thick layer of sediment and made fertile cropland for growing sugar cane. As part of an effort to drain the Everglades, the Mellaleuca tree was imported from Australia. These fast-growing trees are water hungry, meaning that a grove of Mellaleuca can rapidly uptake enough water to dry out a wetland area. Now that we have realized the ecological importance of wetlands like those of the Everglades, we are finding that the Mellaleuca is extremely resistant to our efforts to remove it. In fact, it continues to spread and contribute to habitat loss throughout south Florida.

v. Ornamental species gone wild. We sometimes import plants with beautiful flowers, foliage, or pleasing scents to be used as an ornamental species in landscaping. Some of our most invasive exotic pests are ornamental plants that have escaped into wild ecosystems. Examples include Brazilian pepper, heavenly bamboo, and water hyacinth.

vi. Unwanted or escaped pets. Sometimes when we buy a pet, we do not fully grasp that this cute little baby iguana will one day be three feet long and need a cage the size of a bedroom. We often buy pets such as reptiles, fish, and exotic parrots on impulse. Then after a few months or years, we tire of them. A common thought is to set it free by releasing the pet outdoors. If the climate is right for that animal, it may become established as part of an exotic population. Throughout the southern United States, wild populations of parrots, iguanas, and pythons exist. Most of these animals were released by irresponsible owners or they escaped their captive homes.

D. Efforts to Protect Endangered Species

1. Habitat Protection

   i. The most effective method of protecting endangered species is to preserve their habitat. The benefit of this action is that, by protecting the habitat of a target species, we also make life possible for all the other species that share that habitat.

   ii. Some examples of habitat protection are designating an area as a park or wilderness area, building wildlife corridors between habitat islands, and reducing pollutants from entering the habitat through regulation of local industry and citizens.

2. Captive Breeding Programs

   i. Some species have become so close to extinction or have so little habitat left, that just managing their wild populations may not be enough.

   ii. African cheetah, giant pandas, and some species of tigers are among the critically endangered species being bred in captivity in an attempt to diversify their gene pool and provide a refuge for them while new habitats are being set aside and recovered.

Test Tip

Remember that endangered species are listed by species. For example, do not refer to sea turtles being endangered. Discuss instead leatherback sea turtles, green sea turtles, or hawksbill sea turtles. There are several different sea turtle species and not all of them are endangered.

III. Types of Extinction

A. Extinction leads to the loss of a species. With every extinct species, biodiversity is reduced. Extinctions are classified by magnitude.

1. Mass Extinction

   i.   *Mass extinction* is defined as a series of events causing several different types of species to go extinct in a short period of time.

   ii. To be considered a mass extinction, the event must span several regions of the globe.

   iii. There have been five documented mass extinctions in Earth's history. The most recent was the Cretaceous Extinction 65 million years ago in which all of the non-avian dinosaurs went extinct (along with several other types of animal and plant species).

2. Background Extinction

   i.   Extinctions are not always catastrophic or even caused by humans. Climate changes and natural evolutionary processes cause some species to go extinct naturally.

   ii. The rate at which this happens is estimated to be about one extinction per million species per year. Current estimates put this at about 10 species per year. This is called *background extinction*.

3. Anthropogenic Extinction

   i.   Many scientists speculate that the elevated rate of species being lost due to human activities qualifies our current era as the sixth mass extinction.

IV. Legislation to Preserve Biodiversity

A. Endangered Species Act (ESA)

1. Passed in 1973, the ESA is a far-reaching act that provides protection for any species that is determined to be threat-

ened or endangered with extinction. The provisions of the ESA include strict enforcement of habitat protection, a ban on any activity that will disturb or endanger the life of a listed species, and a ban on the import or export of any individual organisms or product derived from an endangered species.

B. Convention on International Trade in Endangered Species (CITES)

   1. This international agreement regulates the transport of living organisms or products made from any organism on the so-called red list as being endangered. The goal of the convention is to ensure that any international trade does not contribute to the extinction of any animal or plant species.

You are most likely to see the concepts of biodiversity and extinction as a free-response question, in which you will be asked to choose an endangered species and discuss it. Make sure you are very familiar with at least one species that is listed as endangered. Know the full species name, why it is endangered, and what is being done under the ESA to protect it.

REFERENCES

Encyclopedia Britannica Advocacy for Animals: Traditional Chinese Medicine and Endangered Animals. *http:// advocacy.britannica.com/blog/advocacy/2007/10/ traditional-chinese-medicine-and-endangered-animals/*

Convention on International Trade in Endangered Species of Wild Fauna and Flora: What is CITES? *http://www.cites.org/ eng/disc/what.shtml*

PART IV

TYING IT ALL Together

Relating Economics and Environmental Issues

AUTHOR'S NOTE: *Although few of the topics in this chapter relate directly to questions on the exam, there is definitely a strong undercurrent throughout the entire exam that ties together the economy and the environment. Several AP Environmental Science test-takers report that many of the questions required common sense and did not relate specifically to any one concept they had been taught. This common sense approach appears to be the result of grasping the idea that essentially many environmental decisions are made based directly on the underlying economic cost. The aim of this chapter is to get you thinking about why we make decisions that, on the surface, seem not to make any sense at all.*

I. Cost-Benefit Analysis

A. A *cost-benefit analysis* is a way of analyzing the possible options, weighing the costs against the possible benefits, and making the decision that will ultimately bring the greatest benefit. Almost every decision we make uses the cost-benefit analysis process, whether or not we are aware of using it.

    i. In a business setting, the costs and benefits are usually quantified in dollar amounts. Although it is easy to quantify costs in terms of how much a pollution control device or a cleanup operation will cost, other costs, like long-term health effects resulting from exposure to pollution, are more difficult.

    ii. The same applies to benefits. Earned profit is easy, but how much is clean air and water worth? What is the dollar value per acre of high biodiversity? Until we find standardized ways to quantify environmental costs and benefits, they will not be considered in the final equations used by corporate decision makers.

B. Externalized Costs

1. *Externalized costs* are the negative effects of an economic transaction that are not reflected in the actual cost of a product or service. When you buy an inexpensive pair of cotton pants from a store, you are not paying the full cost of the product. Here are just a few of the many externalized costs rarely considered as we are buying those pants:

   i. Pollution caused by pesticides in the cotton fields, and eutrophication from fertilizer runoff into local waterways where cotton is produced.

   ii. Human suffering resulting from deplorable factory working conditions and factory pollutants released in countries with little or no environmental regulation.

   iii. The fossil fuels used to power farming equipment on the cotton farms, to transport the raw materials to the factory, and to transport the finished product to the consumer. Often these inexpensive products have traveled all over the globe by the time they make it to your local store.

C. Full Cost Pricing

1. *Full cost pricing* is the economic approach of including the environmental and human costs in the price of goods and services.

D. Environmental Legislation

1. *Environmental legislation* is one step in the right direction. By imposing fees and fines for pollution and environmental degradation, corporations are forced to add these factors into the cost-benefit equation.

II. Short-Term Versus Long-Term Thinking

A. Short-term thinking drives us to make decisions that will bring the most immediate profit. In most cases, short-term thinking leads to long-term economic loss. For example:

1. If I own a lumber company and lease a stand of old-growth timber, clear-cutting that forest will earn the most money in the short term. Old-growth trees produce high-quality wood in large quantities per tree. It is easier and more efficient to cut all the trees at once, process all of them through the lumber mill, and find a buyer for a large lot of lumber. The question is: What's next? There are no trees left to cut and a new stand of trees needs to be found. Eventually there are no trees left and I am out of business. Not only that, but an entire ecosystem has been destroyed—habitats for non-commercial plants and animals have been taken; soil is more likely to erode and pollute waterways; and frankly, a clear-cut patch of land can be ugly.

B. Long-term thinking drives us to consider sustainable management of resources and allows us to continue making a profit for years to come. Although the initial gain may be less, over the long term, a much greater profit is realized. Let's look at the lumber operation example from above with long-term goals:

1. Same owner, same stand of old-growth trees. Now, however, I am thinking about the long-term sustainability of my lumber operation. Practices like selective logging or thinning smaller trees require more planning and yield a more modest, immediate return. But there are benefits to this kind of management. Because fewer trees are harvested, consumer demand for the old-growth lumber increases, leading to a higher profit per board-foot of lumber. If all the trees are cut down, flooding the market with the lumber will most likely drive prices down.

2. Selective logging and thinning may actually stimulate the growth of forest trees. This means that there will ultimately be more lumber produced on this plot over time than if the forest was untouched. I am also able to satisfy my long-term commercial needs, keep my workers employed, and leave an ecosystem intact, thus protecting soil and water quality while still making a profit.

C. Why Do We Cling to Old, Inefficient Ways of Doing Things?

1. To continue their operations, most publicly held corporations must show their shareholders that they are making a profit. Statements to shareholders are released quarterly—four times per year. As mentioned earlier, long-term thinking does not always yield an immediate profit. It often costs more initially to make sustainable decisions. Here are some examples to consider:

   i. A solid waste management company knows that opening a new landfill in its city will be costly and controversial. Their current landfill is at about 25 percent capacity and is expected to remain in operation for at least twenty more years at current rates of waste. Starting a recycling operation, which would include composting all organic materials, could more than double the lifetime of the landfill. The cost of the recycling facility and the employees needed to sort and process the waste would almost equal the cost of a new landfill. What do you think the waste management company will decide?

   ii. There is a growing number of people trying to generate electricity at home by using solar panels on their rooftops or by installing neighborhood wind turbines. The initial expense of a program like this may be in the tens of thousands of dollars per household. Once the program is installed, however, homeowners no longer have to pay electricity bills. In some cases, they are actually able to sell excess power to the utility, thus making a small profit. In ten years or less, the homeowners have recovered the initial investment and actually begin to make money above and beyond what was spent. Why isn't everyone installing solar panels?

D. The Economics of Prevention Versus Recovery

1. Another aspect to long- versus short-term thinking relates to the difficulty of cleaning up a mess versus the cost of avoiding the mess in the first place.

2. Pollution prevention is always less costly in the long run than pollution cleanup. This is compounded by the fact that, in most cases, even after vast amounts of money have been spent to remediate a polluted area, it is still not going to be as healthy or productive as it was prior to the release of the pollutant.

3. In the past, many corporations were not held financially responsible for pollutants they released into the environment, so they did not design their factories and power plants to include pollution reduction measures.

4. Since the flurry of national and international legislation starting in the 1960s, corporations have been forced to spend millions, even billions, of dollars fixing environmental problems that their pollution has caused. There are now fines for emissions and effluent leaks, which cut into profits. This has led to businesses realizing that reducing pollution at the source is the best economic decision.

III. The Globalization of Money

A. Another important factor to consider is the fact that no nation operates independently of the global market. As discussed earlier, most of the products we buy involve the efforts of people in several countries to produce the raw materials, transport them to factories for production, and bring them to the point of consumption.

B. As long as there are gross differences in the environmental regulations and employee wages of individual nations, we will continue to support unsustainable practices because they are more profitable in the short term.

1. The United States has a minimum wage that must be paid to every worker. This wage is set by the government as a living wage, meaning that it is enough to ensure that a worker has the minimum amount of money necessary to afford housing, food, and the other necessities of life.

2. Many nations, like China, have no minimum wage, and businesses in those countries pay their workers so little that they must provide dorm-like factory housing and feeding programs for their employees. As long as this disparity exists, companies will move manufacturing to China (and other low-wage countries) because goods can be produced for a much lower cost. This ultimately hurts everyone.

    i. In the country of production, workers are forced to work for slave-wages because there is no incentive for companies to pay workers more. Their main goal is to maintain the most competitive pricing.

    ii. Because of the lack of environmental regulation, the pollution produced by factories and other industrial operations remains unchecked and, over time, the ability of these nations to produce raw materials and healthy living conditions for their citizens is greatly reduced.

    iii. Jobs are lost from economies like the United States because manufacturing processes are moved to less-regulated nations, so our unemployment rate rises and our standard of living drops.

3. We know that pollution does not stay near its point of origin. Harmful gases released in the atmosphere rapidly mix with moving air currents and are carried around the globe.

    i. After the Japanese Fukushima Diachi nuclear meltdown occurred in 2011, detectable radiation from the meltdown was measured in the air in the United States within a few months.

    ii. Travel to the most remote, pristine tropical beach in the world and you are likely to find plastic trash washing up on the beaches. Much of the plastic we dispose of ultimately makes its way into the oceans, where it moves worldwide via ocean currents. Once-pristine beaches in remote villages of Costa Rica and Nicaragua are now littered with plastic trash from all over Central America. Hawaiian beaches routinely become awash in trash from countries throughout the Pacific, including China and Japan.

IV. The Wealthy Minority Versus the Rest of the World

A. Underlying Causes

1. The underlying causes of this cycle of environmental destruction can be boiled down to two factors: consumer culture and an ever-increasing wealth gap between the richest people and the poorest people on Earth.

B. Consumer Culture

1. When most people think of Western culture, they think of all the possessions we have (televisions, cars, cell phones, big houses, etc.). As more and more nations increase their level of economic development, many of them are shifting their cultural priorities away from more traditional ways of thinking and toward the more Western consumer culture.

2. We are constantly bombarded with messages from the media, and even our government, that we should be buying more material items and spending more money to keep our economy healthy.

3. Ultimately, this pattern of consumption will lead to exhaustion of the world's resources. A growing number of people disagree with the consumerist philosophy and are attempting to simplify their consumption patterns by buying fewer, high-quality items; looking for fair-trade items (that ensure workers were paid a fair wage); and making more sustainable choices on what foods they eat.

C. The Wealth Gap

1. The wealth gap is the difference between the wealthiest people and the poorest people. We can look at the wealth

gap on a local scale (for example, within a city), on a national scale, or even on a global scale. The truth is, in all of these situations, the wealth gap is growing wider and wider as the wealthiest people become more and more wealthy and the poor slip further into poverty. As long as there are extremely poor people, they will continue to be forced into making environmentally unsustainable decisions to ensure their own day-to-day survival. Only by reducing the wealth gap can we hope to bring all the world's people toward more sustainable living.

The Most Important Environmental Legislation

Many of these acts and treaties have been discussed in other chapters of this book. The intention of this chapter is to give you a quick reference to all of the legislation that is most likely to show up on the AP exam. With the exception of the Kyoto Protocol, all of these acts and treaties have appeared on either the multiple-choice or free-response questions of all released exams. Be sure to familiarize yourself with these acts and treaties and don't spend too much time on others you may have learned along the way.

I. National Environmental Policy Act (NEPA) (1970)

A. One of the first pieces of environmental legislation in the United States, NEPA was written to promote the idea of sustainability.

    1. Environmental Impact Statement (EIS)

        i. Any major federal agency must submit an EIS for any activity that may have a harmful impact on the environment. The agency must outline all possible environmental effects, steps it has taken to avoid environmental harm, and justification for why any unavoidable harm may be necessary. All EISs are reviewed and revised, and must be approved before any proposed project may proceed.

    2. Environmental Protection Agency (EPA)

        i. Soon after NEPA was set into law, President Nixon realized that a federal agency would be needed to implement its goals. Later in the same year (1970), the Environmental Protection Agency was founded. Its primary purpose is to protect human health and the environment and enforces standards under a variety of state and local environmental laws.

II. Water Quality

A. In 1969, the Cuyahoga River in Cleveland, Ohio, caught fire and ignited national outrage at the level of pollution in the nation's waterways. Soon after, pressure was placed on government leaders to pass binding legislation to reduce water pollution.

1. Clean Water Act (CWA)

   i.   Originally passed in 1972 and revised several times since, the mission of the Clean Water Act is to restore and maintain the chemical, physical, and biological integrity of America's waterways to support wildlife and recreation activities. The provisions of the CWA are intended to reduce and prevent both point and non–point sources of water pollution in the nation's surface waters. There are no provisions specific to groundwater protection.

2. Safe Drinking Water Act (SDWA) of 1974

   i.   The focus of the Safe Drinking Water Act is to maintain the purity of any water source that may potentially be used as drinking water. This includes both surface water and groundwater.

III. Air Quality

Test Tip

The only piece of U.S. legislation related to air quality that you need to know for the exam is the Clean Air Act. The Montreal and Kyoto Protocols are also significant because each deals with a major environmental concern.

A. The Clean Air Act

1. Beginning in 1955, the need became evident for some type of national legislation to reduce air pollution. Regulations were passed that in 1963 became the Clean Air Act. This legislation has been modified several times throughout the years. The provisions of the act include the following:

    i.   Authorization and funding to study air quality in order to learn more about the presence and effects of atmospheric pollutants.

    ii. Setting enforceable regulations to limit emissions from stationary sources (factories and power plants) as well as mobile sources (cars, trucks, and ships).

    iii. Developing programs to monitor and reduce acid deposition and the primary pollutants that cause its formation.

    iv. Establishing a program to phase out the use of chemicals that deplete stratospheric ozone.

    v.   Establishing a cap and trade program for SO_2:

        ➤ *Cap and trade* is a system in which maximum allowable emissions are set for each industry.

        ➤ Businesses that can reduce their emissions below the standards are awarded credits. They can then sell their credits to other businesses that cannot meet the limits.

        ➤ This makes economic sense because it creates a huge financial incentive for industry to reduce emissions quickly. Allowing business firms to sell emissions credits means that many groups can rapidly recover the costs associated with reducing emissions.

B. Montreal Protocol (1987)

1. One of the greatest environmental success stories to date, the Montreal Protocol is proof that global problems can be solved by a combination of international cooperation and scientific advances. The provisions of the Montreal Protocol

require participating nations to phase out the use of ozone-depleting chemicals in favor of less harmful alternatives.

C. Kyoto Protocol

1. The Kyoto Protocol was developed at a world summit in 1997 to address measures for reducing greenhouse gas emissions. Many promises were made by many nations, but no real progress was made. The United States announced that it would not sign the protocol, and soon after, the whole movement lost support.

IV. Solid and Hazardous Wastes

A. Comprehensive Environmental Response, Compensation, and Reliability Act (CERCLA) (1980)

1. Commonly known as the Superfund Act, CERCLA was enacted to handle industrial contamination in sites where no direct individual or party could be held responsible for cleanup.

2. The law imposed a tax on the chemical and petroleum industries and authorized the federal government to respond to the release of hazardous substances that might endanger public health or the environment.

3. The act also created a trust fund for cleaning hazardous waste sites when no responsible party could be identified.

B. Resource Conservation and Recovery Act (RCRA) (1976)

1. Commonly called the "Cradle-to-Grave Act," this legislation sets specific regulations concerning the manufacture, transport, storage, use, and ultimate disposal of a host of hazardous chemicals.

2. Its major provision requires extensive documentation at every step to ensure that hazardous wastes are disposed of properly.

V. **Endangered Species Act (ESA)**

1. Passed in 1973, the ESA is a far-reaching act that provides protection for any species that is determined to be threatened or endangered with extinction.

2. The provisions of the ESA include strict enforcement of habitat protection, a ban on any activity that disturbs or endangers the life of a listed species, and a ban on the import or export of any individual organisms or product derived from an endangered species.

REFERENCES

U.S. Fish and Wildlife Service Digest of Federal Resource Laws. *http://www.fws.gov/laws/luwsdigest/resourcelaws.htm*

PART V

TEST-TAKING
Strategies

Strategies for the Multiple-Choice Questions

When you sit down on test day and open your AP test booklet, you should already have your plan of action in place. Working your way through the 100 questions on the multiple-choice section of this test may seem grueling at times, but if you have a strategy in mind, you will be able to keep your momentum going.

On test day, you need, above all, to keep a positive attitude. If you walk into the test well rested and confident that you are prepared, you have put yourself in an ideal position for success. To maintain that positive attitude, remember that you *do not need a perfect score to earn a 5* on this exam. In fact, if your goal is to pass with just a 3, you only need to earn about 45 percent of the total points on the exam. If your goal is a 4 or a 5, then aim for 60 to 70 percent of the total points. This means that you can miss almost a third of the questions and still earn a 4 or a 5!

1. START WITH THE EASY QUESTIONS

You will have 90 minutes to answer 100 multiple-choice questions. That allows you less than one minute to answer a question. Bring an easy-to-read watch with you on exam day (but make sure it's not a calculator watch; calculators are not allowed). Begin by reading through the exam questions and answering any questions you are sure you know. Circle any questions that will require a little more thought and come back to them later. Remember, you are collecting points toward your goal. As long as you answer about half or more of the questions correctly on this first pass, you will have accumulated enough points to earn a 3 without even tackling the difficult questions. Now, check your time. You should have 30 minutes or more to go back and try the harder questions.

2. TO GUESS OR NOT TO GUESS?

There is no guessing penalty on AP exams. This means that you absolutely do not want to leave any questions unanswered. On your second pass through the questions, eliminate answer choices that you know are incorrect, and then guess (if necessary) from the choices that are left. For every incorrect answer choice you can eliminate, you greatly increase your chances of guessing right. Be sure to read the whole question carefully and pay attention to the wording of each.

3. TYPES OF MULTIPLE-CHOICE QUESTIONS

The multiple-choice questions are divided into two sections, Part A and Part B.

Part A (usually the first 15 to 20 questions) is set up as a group of matching questions. You will be given a list of five choices (lettered (A) through (E)), and then three to five questions that relate to those choices.

Common topics covered in the matching questions include the following:

➤ Lists of chemical formulas you should know (NO_x, SO_x, H_2O, CO_2, N_2, CH_4, CCl_2F_2) or names of chemical pollutants.

➤ Country names related to population characteristics or specific resource reserves.

➤ Titles of legislation.

➤ Layers of the atmosphere.

➤ A list of related vocabulary terms.

The questions in Part A can also relate to a map. Two of the three released exams started with a map.

The map might show a view of all the continents, but there are no labels other than the letter choices. Be sure you can identify all of the continents and major countries like the United States, Canada, Brazil, India, China, Japan, Australia, Antarctica, and Russia. Also review major geographical regions like western Europe, Central America, Asia, the Middle East, and so on.

The questions often relate to identification of biomes; areas of coal, oil, or natural gas reserves; population growth characteristics; or areas of tectonic activity.

Part B makes up the remainder of the multiple-choice questions. These questions are all similar in that they have a statement or a question followed by five answer choices. In Part B, you will find general knowledge questions, as well as calculations, data interpretation, and the multiple-answer questions.

CALCULATION QUESTIONS

There will be about four to six calculation questions in the multiple-choice section. Because you are not allowed to use a calculator, these are generally more about whether you know how to set up the problem rather than the actual calculations. If the numbers look like they will be difficult to work with, round up or down to whole numbers and estimate an answer. For the most part, your answer choices will be something like 1, 10, 100, 1,000 or 10,000, so estimating will work just fine. If you have no idea, try working backward by plugging one of the answer choices into your equation. If you are running short on time, guess at the answer and save your brain power for the remaining questions. If math is not your strong point, remember that you can miss every math question and still earn a 4 (and possibly a 5) on the exam. Don't let math trouble you!

DATA INTERPRETATION QUESTIONS

These questions are like free points. You don't really need to remember much in terms of factual information; you just need to know how to read the data off a graph or chart, and compare numbers to one another. There will usually be three or four questions of this type on your exam.

(THE EVIL!) MULTIPLE-ANSWER QUESTIONS

These questions are truly a hassle, but you will see only about three of them. A multiple-answer question gives you a list of statements numbered I, II, III, and IV. It then may ask which statements are true, which factors contribute to a particular problem, or even which are good solutions for a particular problem. Your answer choices will look something like this:

A. I only

B. II only

C. II and III

D. I, II, and III

E. I, III, and IV

Sometimes, these types of questions actually end up being pretty simple because you can easily eliminate one or more of the numbered statements, but other times you just need to guess and move on.

4. TRUST YOUR INSTINCTS

Most important is that you don't lose your confidence and start second-guessing yourself. If you can't remember the answer to a question or two, do NOT go back and start changing your answers. It is a good idea, however, to review your test to check that you have not made any careless errors while filling out the answer sheet. Do not change any answers unless you are absolutely certain that your original choice was clearly a mistake. Research shows that, in the vast majority of cases, your first instinct is more likely to be correct than your second guess. So relax, do your best, and don't worry about the rest!

Strategies for the Free-Response Questions

On the day you take the AP Environmental Science exam, you will begin with the 90-minute multiple-choice session. Once that session is over, you will be given a 15-minute break before the free-response portion begins. I suggest that you take the break as an opportunity to get outside if possible. Stretch your muscles, especially your shoulders and back, to release any tension that may have built up during the last 90 minutes. Breathe some fresh air. Have a snack. At some schools, snacks and drinks are provided for break times; find out if you will be given a snack, or be safe and bring your own. Avoid talking with others about the test. This is a break to clear your mind—don't do anything that will increase your stress level.

1. PACE YOURSELF

When it is time to take the free-response portion of the exam, you will have 90 minutes to answer four questions. Each question will be worth ten points and have three to five parts that all relate to a common theme. Decide ahead of time how you want to allocate your time. Make a schedule and stick to it. The No. 1 complaint from poor-performing students is that they lose track of time. The exam proctor will not make announcements telling you how much time you have left, so it is up to you to keep track for yourself. This is where your watch really is essential. Here are two strategies for dividing your time.

A. TEN-MINUTE READING PERIOD WITH 20 MINUTES PER QUESTION

This is a great strategy for most students and the one I recommend you use.

TEN-MINUTE PLANNING TIME

You will be given two booklets, a green planning book and a pink answer book. With this strategy, take your first ten minutes to read each question closely.

Use this time to decide which question you will answer first, second, third, and fourth. You do not need to answer the questions in any specific order, just be sure you *clearly* mark what number and letter of the question you are answering for each part of the four questions.

Use the green planning booklet to jot down key terms you want to remember to use in your answer. Draw a star by parts of a question that you can easily answer, and circle parts that may be more difficult or take more time.

Watch the clock! You have about 2 minutes per question to read and plan. When your 10 minutes are up, it's time to get down to business!

TWENTY MINUTES TO ANSWER EACH QUESTION

I recommend that you start with the question next to which you drew the most stars. If you have starred all the easy parts, you will see quickly which question will be the best to answer first.

Remember that you are accumulating points toward your goal, and you don't need them all. Answer the parts of the question that are the simplest and save the parts that require more thought for later. If you pace yourself, you will have enough time to do everything, but if you run out of time, at least you answered the parts that are most likely to be correct and earn you more points.

Keep your eye on the clock. When your 20 minutes are up, move on even if you haven't finished your answer. Just leave enough space so you can go back to finish your answer later. Chances are, you will use less than 20 minutes on one or more of the other questions, so you will have some time at the end to tie up loose ends.

B. TWENTY-TWO MINUTES PER QUESTION WITH NO READING PERIOD

Some people are just linear thinkers. If the thought of answering the questions out of numerical order sounds absurd to you, then you may want to jump right in instead of reading over all the questions first.

Even with this approach, I recommend that you at least take a quick glance through all four questions to get an idea of which ones you'll find easiest.

Use the first 2 minutes for each question to read all parts of the question and plan your approach for answering each part. Use the green book to jot down key terms or ideas you have while reading the question.

2. GENERAL STRATEGIES FOR ANSWERING QUESTIONS

Clearly label the question number and each part of the answer. Skip a line or two between parts to make it easier for the AP grader to find your answers. This also gives you some space in case you remember something later that you want to add.

Do not waste time restating the question in your answer or writing an introductory sentence—this is not an English test.

Keep your answers short, simple, and to the point. Use complete sentences, but do not fill them with flowery language.

If the question states, "Discuss the economic advantages of solar power," do not start your answer by writing "Solar power has many economic advantages over other sources of energy." Jump right in and start writing about solar power. The test reader knows the question—save your time and space for details that will earn you points. Don't waste time on fluff!

Be aware that, in many cases, the most difficult part of a question is the beginning. If you can't figure out what to write for the first and or second parts of a question, *do not give up!* The last two or three parts of a question are often the easiest and worth the most points. Skip down and answer the last parts first. Then, if you have time, go back and tackle the tricky part at the beginning.

Be specific! Avoid using words like *many* and *various* and phrases like *certain kinds*. If there are various air pollutants, list specific ones. If certain kinds of chemicals cause cancer, describe them. A vague answer will probably earn very few points. Words like *toxic, pollution, harmful,* and *unsustainable* are good starting points, but you must elaborate with specific information to earn points. This is your chance to show how much you know, so put it all down in as much detail as you possibly can.

3. TERMINOLOGY USED IN FREE-RESPONSE QUESTIONS

As you read through each question, pay special attention to the way the question is asked. Below is a list of the most common terms you might see in a question and how to address them.

DISCUSS/DESCRIBE/EXPLAIN

These are the three most common terms used in questions. Start by defining any terms you will be using in your answer, then write a sentence or two adding details. Be sure to include at least one or more examples to back up your description.

COMPARE AND CONTRAST

Compare means to describe all the details that your two topics have in common. *Contrast* means to show all of the distinct differences between the topics. Be sure to include both in your answer. Once again, use specific examples whenever possible.

JUSTIFY

Discuss your reasons for making a choice. In many cases, it will be appropriate to tie your answer to which choice is more environmentally sustainable. Give specific reasons that support your decision and include examples where appropriate.

LIST

Simply name one or more items that fit the description—no further description needed. Beware! Many questions ask you to "list and describe." In this case be sure to give a detailed description for each item you list.

DESIGN AN EXPERIMENT

It is possible that you will be asked to design an experiment as part of your answer. About 7 percent of all past free-response questions have included experimental design in one or more parts of the question. Make sure you include all of the essential elements to a valid experiment:

1. Clearly state your hypothesis or predicted outcome.

2. Describe your test groups and your control group.

3. Manipulate only one variable at a time. (For example, note a change in temperature only; not a change in both temperature and light.)

4. List all of the factors that will be constants for all groups.

5. Describe how you will collect your data. Be specific.

6. Discuss what results and conclusions you expect to see, based on your knowledge of the underlying concepts.

7. Remember to mention the importance of replication to increase the validity of your data.

4. TYPES OF FREE-RESPONSE QUESTIONS

DOCUMENT-BASED QUESTIONS (DBQs)

AP Environmental Science DBQs are much more simple than the DBQs you will see on the AP U.S History exam. In most cases, your first question will start with a simulated clipping from a newspaper article, interview, or other publication. The document will provide some background information on a topic you have studied and will often include one or more opinions on the topic. You will then be asked to do some critical thinking related to the theme of the article.

Read the questions carefully. In many cases, you will be asked to pull information directly from the article for one or more parts of the question. Make it obvious when you do this. For example, you could write, "According to Dr. Tate's statement in the article . . . ."

The final parts of the question will require you to add what you already know about the topic. You may be asked to give an additional example of a similar situation, or you may be asked to propose and justify a solution to a problem.

Most people find this question to be the easiest of the four free-response questions.

SYNTHESIS AND KNOWLEDGE QUESTIONS

One or more of your questions will be a synthesis and knowledge question. Generally, this type of question begins with a sentence or two describing a situation or environmental issue. Each part of the question somehow relates back to the theme of the question.

For example, if the theme of the question is endangered species, the question may begin by discussing the giant panda and its habitat. It may go on to ask you to describe an endangered species in the United States, discuss traits that make a species vulnerable to extinction, and finally ask for a description of legislation enacted to protect endangered species.

DATA- OR CALCULATION-BASED QUESTIONS

Usually two of your questions will in some way involve data interpretation and/or mathematic calculations. Pay close attention to the wording of the question and be sure to follow the directions exactly!

Data interpretation questions present you with a data set in the form of a table, map, graph, or diagram. You will then be asked to interpret that data. You may need to describe the parts of a diagram; make a graph from a set of data; draw a food web; or use numbers in a diagram, graph, or table to perform a calculation.

Calculation questions require that you show your work. Make sure you set up a logical progression of your calculation(s), and include units in your initial setup and your answer.

In most cases, a calculation question is worth 2 points: 1 point for the setup and 1 point for the answer. You can get a setup point even if you make a calculation error and get the wrong answer. If you don't show your work, however, chances are good that you won't earn any points!

If you are not good at math, at least set up the logical progression you would follow to do the calculation. Even if you forget how to do long division, you can earn the setup point without attempting the calculation, as long as your logic is correct and you include the appropriate units.

If you have a math question that requires the answer from part (a) to do the calculation in part (b), you must use the answer you calculated in part (a), or no points will be awarded. If you have no idea how to do part (a), guess a number, write it as your answer for part (a), and proceed to use it in other parts of the question where required. You will not earn the points for part (a), but you will not be penalized in later parts of the question as long as those calculations are done correctly using your value for part (a).

In many cases, the calculation-based question requires calculations only on the first few parts. The remaining parts of the question are often easy. If you have no idea how to do the calculations, be sure to skip down and do the later parts of the question that do not require using the numbers. In many cases you can get four or five of the ten points without even attempting to do the calculations!